Henry T Bovey

The Strength of Canadian Douglas Fir, Red Pine, White Pine, and Spruce

Henry T Bovey

The Strength of Canadian Douglas Fir, Red Pine, White Pine, and Spruce

ISBN/EAN: 9783337205584

Printed in Europe, USA, Canada, Australia, Japan

Cover: Foto ©Andreas Hilbeck / pixelio.de

More available books at **www.hansebooks.com**

THE STRENGTH OF CANADIAN DOUGLAS FIR, RED PINE, WHITE PINE AND SPRUCE.

PROFESSOR H. T. BOVEY, LL.D., F.R.S.C.

THE STRENGTH OF
CANADIAN DOUGLAS FIR,
RED PINE, WHITE PINE AND SPRUCE.

By Henry T. Bovey, M.Inst.C.E., LL.D.
(Read January 25, 1895.)

In the present Paper it is proposed to give a statement of the results which have been obtained up to the present time, from the numerous experiments which have been carried out in the Testing Laboratories, McGill University, on the strength of Canadian Douglas Fir, Red Pine White Pine and Spruce.

These experiments, which have now extended over a period of more than two years, will still be continued, and it is hoped that the results, will be set before the profession in a Paper on some future occasion.

In order that the subject may be treated in as comprehensive a manner as possible, the engineers and lumber merchants, who must necessarily be most particularly interested, are earnestly requested to give their co-operation. They can render valuable service by sending to the University Laboratories timbers of any and all sizes. These timbers should, in each case, be accompanied by a history giving the treatment of the timber from the time when the tree was felled, as, for example, the locality in which the tree grew should be specified, the manner in which the log was brought to the mill, the length of time during which it was kept in water (salt or fresh), the time during which it was kept in the pile at the mill, and, if the timber has already been in service, the length of this service. Any other details respecting the history of the timber may also be given, so that the information may in every case be as complete as circumstances will permit.

The attention of members is specially directed to the tables showing the deflection of beams under transverse loading, and also to tables showing the extension of specimens under direct tension.

These tables tend to prove conclusively the statement made by the author many years ago, i.e., that timber, unlike iron and steel, may be strained to a point near the breaking point without being seriously injured. It will be observed that in almost all cases the increments of deflection and extension, almost up to the point of fracture, are very nearly proportional to the increments of load, and it seems impossible to define a limit of elasticity for timber. This probably accounts for the continued existence of many timber structures in which the timbers have been and are still continually subjected to excessive stresses, the factor of safety being often less than $1\frac{1}{2}$. Whether it is advisable so to strain timber is another question, and experiments are still required to show how timber is affected by frequently repeated strains.

TRANSVERSE STRENGTH.

The following Table gives in inches the distances between the centres of the end bearings (l), the mean depths (d) and the mean breadths (b) of the Beams I to LXI referred to in this Paper:—

Beams	I	II	III	IV	V	VI	VII
l	96	66	66	69	69	69	69
	×	×	×	×	×	×	×
d	12.125	12.125	5.375	9.125	9.125	6.125	6
	×	×	×	×	×	×	×
b	9	5.625	4.25	5	5	6	5.3125

Beams	VIII	IX	X	XI	XII	XIII	XIV
l	69	204	198	204	204	204	204
d	5.125	14.875	14.875	14.875	14.875	14.75	14.75
b	5.5	9	6	8.6875	8.8125	6	6
Beams	XV	XVI	XVII	XVIII	XIX	XX	XXI
l	198	198	138	138	138	138	138
d	15	15	15.125	17.8	12.1	12	8.98
b	6.125	6.125	9	8.76	9.1	8.88	5.95
Beams	XXII	XXIII	XXIV	XXV	XXVI	XXVII	XXVIII
l	162	186	132	114	210	210	210
d	15.6875	14.35	16.2	15.65	13.25	13.125	11.25
b	7.75	8.78	7.75	8.2	6.375	6.1875	6.34375
Beams	XXIX	XXX	XXXI	XXXII	XXXIII	XXXIV	XXXV
l	210	174	174	180	180	156	156
d	11.25	7.25	7.125	8.125	11.125	9.125	11.15
b	6.25	6.1875	6.21875	3.1	3.1	3.125	3.325
Beams	XXXVI	XXXVII	XXXVIII	XXXIX	XL	XLI	XLII
l	288	288	114	102	120	120	288
d	18	18	18	18	18	18	18
b	9	9	9	9	9	9	9
Beams	XLIII	XLIV	XLV	XLVI	XLVII	XLVIII	XLIX
l	120	120	288	120	120	150	15
b	18	18	18	18	18	15.1875	15.375
b	9	9	9	9	9	9.375	9.125
Beams	L	LI	LII	LIII	LIV	LV	
l	186	192	180	180	288	120	
d	15	15.12	14.85	15	17.5	17.5	
b	9.0625	9	9.05	9.05	8.875	8.875	
Beams	LVI	LVII	LVIII	LIX	LX	LXI	
l	120	188	180	180	138	186	
d	17.5	15	11.75	15	11.25	11.5	
b	8.9375	9	6	9	8.875	5.625	

The transverse tests were carried out with the Wicksteed 100-to machine by means of a specially designed arrangement shown in the photograph on the opposite page.

By this arrangement the two ends are gradually forced downwards while the centre is supported upon the saddle suspended from the lever of the machine. Thus the two halves of the beam are really equivalent to two cantilevers loaded at the ends. By means of a very simple device, the pressure can be increased so regularly as to ensure an absolute equality in these end loads.

Figures 1 and 2 show the device employed to keep the pressure on the ends of the beam always normal to the surface. The spherical

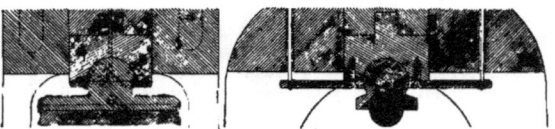

joint allows the bearing to revolve, and by means of the prismatic slot any form of bearing surface may be introduced.

The formula used in calculating the skin-strengths and co-efficients of elasticity have been deduced by means of the ordinary theory of flexure which is based upon assumptions which actual experience shows to be far from being true. These assumptions are :—

(a) That the beam is symmetrical with respect to a certain plane.

(b) That the material of the beam is homogeneous.

(c) That sections which are plane before bending remain plane after bending.

(d) That the ratio of longitudinal stress to the corresponding strain is the ordinary (i. e. Young's) modulus of elasticity, notwithstanding the lateral connection of the elementary layers.

(e) That these elementary layers expand and contract freely under tensile and compressive forces.

In each case, the skin stress at the point of fracture in lbs. per sq. in has been determined by means of the formula,

$$f = \tfrac{3}{2} \frac{l\,(2\,W_1 + W_2)}{b\,d^2}$$

W_1-lbs. being the weight at an end, W_2-lbs. half the weight of the beam l-ins. the length of the beam between the two end centres of pressure, b-ins. the breadth and d-ins. the depth at the section of fracture.

In practice, the breaking weight, $W_1 + \tfrac{1}{2} W_2$, is usually determined from the formula,

$$W_1 + \tfrac{1}{2} W_2 = C\,\frac{b\,d^2}{l},$$

C being the co-efficient of rupture. Hence, $f = 3\,C$.

It may perhaps be well to point out that a very small error in estimating the depth of a beam may lead to a considerable error in the calculated skin stress. Thus from the formula just given it appears that if Δf be the change in the skin stress corresponding to a change Δd in the depth, then

$$\Delta f = -\,2\frac{f}{d}\Delta d,$$

and the skin stress will be increased or diminished by this amount, according as the estimated depth is too small or too great by the amount Δd.

For instance, in the case of the Spruce Beam No. L, the calculated skin stress, disregarding the diminution of depth due to compression, is 5123 lbs. The initial depth (d) of the beam was 17.5 ins., and the amount of the compression (Δd) 2 ins. Thus the error (Δf) in the skin stress is

$$\Delta f = -\,2\frac{5123}{17.5} 2 = 1171 \text{ lbs. per sq. in.},$$

and the actual stress becomes $5123 + 1171 = 6294$ lbs. per sq. in., showing an increase of 22.8 per cent.

Now, in every example of transverse testing, the material is more or less compressed at the central support. The central support in the following examples was a hardwood block of 20 ins. diameter. The amount of the compression at this support depends not only upon the nature of the material of the beam and upon the character of the support, but also very especially upon the ratio of the length of the beam to its depth. In calculating the skin stress corresponding to the breaking weight, therefore, three assumptions may be made :—

1st. That the compression at the support may be disregarded.

2nd. That the effective depth of the beam may be taken as equal to the initial depth minus the amount of the compression, and that the usual law may be assumed to hold good for the whole of this effective depth.

3rd. That the compression portion of the beam is alone affected, so that the so called neutral plane remains in the same position relatively to the tension face of the beam from the commencement of the test to the end.

Calculations based upon these three assumptions have been made in several of the following cases, and it will be observed that in all cases the skin stress calculated upon the first assumption is invariably less than the skin stress determined upon either of the remaining assumptions.

Thus any error is on the safe side.

It should be remembered, however, that it is possible, and even probable, that neither of these assumptions is even approximately correct, at all events, beyond the limit of elasticity, which in the case of timber still remains indefinite. The portion in compression doubtless acquires increased rigidity, and thus exerts a continually increasing resistance, so that there is produced a more or less perfect equalization of stress throughout the portion of the beam under compression, and this equalization will doubtless materially affect both the elasticity and the strength.

An interesting paper on the surface-loading of beams was presented by Prof. C. A. Carus-Wilson to the Physical Society of London, (Eng.), and an abstract of this Paper is to be found in the author's treatise on the Theory of Structures.

The co-efficient of elasticity, as determined by the tranverse loading, is deduced from the formula

$$E = \frac{1}{4} \frac{\Delta W}{\Delta D} \cdot \frac{l^3}{bd^3}$$

W being the increment of weight corresponding to the increment ΔD of the deflection.

Here again an error Δd in the estimated depth will produce an error ΔE in the calculated co-efficient of elasticity measured by

$$\Delta E = -3 \frac{E}{d} \Delta d.$$

DOUGLAS FIR.

Beams I to III were sent to the Testing Laboratory by Mr. John Kennedy, Chief Engineer of the Montreal Harbour Works.

Beams I and II were of good average quality.

Beam I was tested on March 1st, 1893, with the annular rings as in Fig. 3. The load was gradually increased until it amounted to 45,000 lbs., when the beam failed by the tearing apart of the fibres on the tension face.

The maximum skin stress corresponding to the breaking weight of 45,000 lbs. is 4897 lbs. per square inch.

The co-efficient of elasticity, as deduced from an increment in the deflection of .23-in. between the loads of 3500 and 22,500-lbs., is 1,138,900 lbs.

Table A shows the several readings.

Beam II was tested on March 2nd, 1893, with the annular rings running as in Fig. 4.

The load was gradually increased until it amounted to 36,575 lbs. when the beam failed by shearing longitudinally.

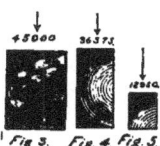

Fig 3.   Fig. 4. Fig. 5.

The maximum skin stress corresponding to this breaking weight is 4378 lbs. per square inch.

In connection with this experiment it is of interest to note that the timber, although it had failed by longitudinal shear, still possessed a very large amount of transverse strength, and similar facts will be subsequently referred to in the case of other beams. After the fracture, the load upon the beam was again gradually increased to 34,000 lbs. before a second failure occurred.

The co-efficient of elasticity, as determined by the increment in the deflection of .1 in. between the loads 2000 and 18,000-lbs., is 1,146,900 lbs.

Table B shows the several readings.

Beam III was tested on March 2nd, 1893, with the annular rings as in Fig. 5.

This Beam was of especially excellent quality, with clear, close, parallel grain, perfectly sound and free from knots.

The load was gradually increased until it amounted to 12,950 lbs., when it failed by shearing longitudinally.

The maximum skin stress corresponding to the breaking load is 10,441 lbs. per square inch.

The co-efficient of elasticity, as determined by an increment in the deflection of .2-in. between the loads of 500 and 4500-lbs., is 2,173,100 lbs.

Table B gives the several readings.

Beams IV to VIII were sent to the laboratory by the British Columbia Mills Timber & Trading Company through Mr. C. M. Beecher.

These beams were cut out of trees grown on the coast section of British Columbia, and felled in the fall or during the winter. The whole of the beams were free from knots, of good quality, and with the grain running straight from end to end.

Beam IV was tested May 17th, 1893, with the annular rings somewhat oblique as shown in Fig. 6. Under a load of 16,720 lbs. it

Figure 6.

failed by shearing longitudinally along a plane AB at right angles to the annular rings, the distance between the ends of the portions above and below the plane of shear being ¼-in. The plane of shear extended to a distance of about 36 ins. from the end of the beam.

The maximum skin stress corresponding to the breaking load is 4156 lbs. per square inch.

The co-efficient of elasticity, as determined by an increase in the deflection of .14-in. between the loads of 2,000 and 8,000 lbs., is 926,500 lbs.

Table B shows the several readings.

After the beam had sheared longitudinally, the jockey weight was run back, and the load again gradually applied until it amounted to 15,000 lbs., when fracture occurred by the tearing apart of the fibres on the tension face. Under this load of 15,000 lbs. an opening of ½-in. was developed in the end at the plane of shear.

On May 11th this beam weighed 56 lbs. 13 ozs., or 28.59 lbs. per cubic foot. On May 17th, the weight of the beam was 56 lbs. 3 ozs., or 28.27 lbs. per cubic foot, so that while in the laboratory this beam lost in weight at the rate of .0533-lb. per cubic foot per day.

Beam V was tested on May 19th, 1893, with the annular rings somewhat oblique as shown in Fig. 7. It failed by the tearing apart of the fibres on the tension face under a load of 23,610 lbs.

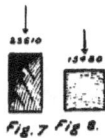

Fig. 7 Fig. 8.

The maximum skin stress corresponding to this load is 5869 lbs per square inch.

The co-efficient of elasticity, as determined by an increase in the deflection of .24-in. between the loads of 1000-lbs. and 11,500-lbs., is 946,270 lbs.

Table B shows the several readings.

The weight of the beam on May 11th was 59 lbs., or 29.59 lbs. per cubic foot. The weight of the beam on May 19th was 58 lbs. 3 ozs., or 29.18 lbs. per cubic foot, so that the loss in weight in the laboratory was at the rate of .05125-lb. per cubic foot per day.

Beam VI was tested May 22nd, 1893, with the annular rings as in Fig. 8. Under a load of 15,480 lbs. it failed by the tearing apart of the fibres on the tension face.

The corresponding maximum skin stress is 7116 lbs.

The co-efficient of elasticity as determined by an increase in the deflection of .3-in. between the loads of 500-lbs. and 8,000-lbs. is 1,489,215 lbs.

Table B shows the several readings.

The weight of the beam on May 11th was 49 lbs. 6 ozs., or 31.05 lbs. per cubic foot, and the weight on May 22nd was 48 lbs. 1 oz., or 30.23 lbs., showing a loss of weight while in the laboratory at the rate of .0745-lb. per cubic foot per day.

Beam VII was tested on May 19th, 1893. In this beam the annular rings ran somewhat obliquely as in Fig. 9. Under a load of 17,615 lbs., the beam sheared longitudinally along the plane AB, Fig. 10, the distance between the ends of the portions above and below the plane of shear being 3-16ths of an inch. The plane of shear extended to a distance of 46-ins. from the end of the beam.

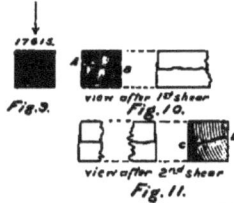

Fig. 9. view after 1st shear Fig. 10.

view after 2nd shear Fig. 11.

The maximum skin stress corresponding to this breaking weight of 17,615 lbs. is 8712 lbs.

The co-efficient of elasticity, as determined by an increase in the deflection of .255-in. between the loads of 500 lbs. and 8500 lbs., is 2,052,250 lbs.

Table B shows the several readings.

Immediately after the longitudinal shear the jockey weight was run back until it indicated a load of 5090 lbs. when the lever again floated. The weight was then gradually increased until it amounted to 11,840 lbs., when there was a second longitudinal shear along the plane CD at the other end, Fig. 11. The lap at the plane AB was now increased from 3-16ths in. to 3-10ths in., and the distance between the ends of the portions above and below the plane of shear at the other end of the beam was 3-20ths of an inch.

After this second shear the jockey weight was run back to 6840 lbs when the lever floated. The load was gradually increased until it amounted to 8990 lbs., when the beam was fractured by the tearing apart of the fibres on the tension face.

On May 11th, this beam weighed 60 lbs. 4 ozs., or 40.69 lbs. per cubic foot, and the weight on May 19th was 59 lbs. 2 ozs., or 39.92 lbs. per cubic foot, showing a loss of weight in the laboratory at the rate of .09625-lb. per cubic foot per day.

Beam VIII was tested May 22nd, 1893. In this beam the annular rings were oblique as in Fig. 12. Under a load of 11,700 lbs. it failed at the support by the tearing apart of the fibres on the tension face.

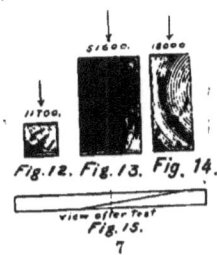

Fig. 12. Fig. 13. Fig. 14.

view after test Fig. 15.

The maximum skin stress due to this load is 8382 lbs. per square inch.

The co-efficient of elasticity, as determined by an increase in the deflection of .32-in. between loads of 1000 lbs. to 5500 lbs., is 1,559,950 lbs.

Table B shows the several readings.

The weight of this beam on May 11th was 44 lbs., or 36.76 lbs. per cubic foot, and its weight on May 22nd was 42 lbs. 14 ozs., or 35.74 lbs. per cubic foot, showing a loss of weight in the laboratory at the rate of .0927-lb. per cubic foot per day.

Beams IX to XVI were sent to the laboratory by Mr. P. A. Peterson, chief engineer of the Canadian Pacific Railway.

Beam IX was grown on the mainland half way between Vancouver and New Westminster, in a flat country not much above the sea level. It was cut from a log 26 ins. in diameter and 34 feet in length, which was felled about the month of May, 1892. The log was floated to the mill at Vancouver, and lay in fresh water for ten months.

The timber corresponded to first quality in the market, its grain being straight and running parallel to the axis. It contained a season crack on the widest face, about 11 feet long, 3½ ins. below the edge, and about 1½ in. deep. The beam was tested Nov. 13th, 1893, with the annular rings as in Fig. 13, the heart of the tree being in one of the vertical faces. Under a load of 51,600 lbs. this beam failed at the support by the tearing apart at the centre of the fibres on the tension face.

The maximum skin stress corresponding to this load is 7974-lbs. per square inch.

The co-efficient of elasticity, as determined by an increment in the deflection of .77-in. between the loads of 1000-lbs. and 20,000-lbs., is 1,767,990 lbs.

Table C shows the several readings.

The weight of the beam was 603 lbs., or 36.49 lbs. per cubic foot on Oct. 3rd, 590 lbs. 13 ozs., or 35.76 lbs. per cubic foot on Nov. 10th, and 590 lbs. on Nov. 13th, showing a loss of weight while in the laboratory at the rate of .0195-lb. per cubic foot per day.

Beam X. This beam was tested Nov. 11th, 1893, with the annular rings as in Fig. 14. It was cut from a log 32 ins. in diameter grown on the mainland 120 miles north and west of Vancouver, on a hill side about 100 feet above the sea-level. The log was felled in the winter of 1892-93, and was then towed to the mill, and remained in salt water six months.

The grain in this beam ran crosswise, and it failed by a cross fracture along the plane AB, Fig. 15.

The fracture occurred under a load of 18,000 lbs., corresponding to a maximum skin stress of 4027 lbs. per square inch. The co-efficient of elasticity, as determined by an increase in the end deflections of .84-in. between the loads 1000-lbs. and 15,000 lbs., is 1,637,806 lbs.

Table C shows the several readings.

The weight of the beam was 407 lbs. 2 ozs., or 38.94 lbs. per cubic foot on Oct. 3rd, 406 lbs. 8 ozs., or 37.80 lbs. per cubic foot on Nov. 10th, and 404 lbs. 13 ozs., or 37.79 lbs. per cubic foot on Nov. 13th, showing a loss of weight in the laboratory at the rate of .03-lbs. per cubic foot per day.

Beam XI. This beam was tested November, 7th, 1893, with the annular rings as in Fig. 16. Its history is the same as that of Beam

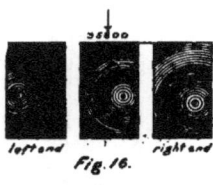

Fig. 16.

X. The timber was of a quality corresponding to first quality in the market, and the grain for the most part was parallel with the axis. It contained a few season cracks. On the tension face of the beam the fibres crossed from back to front in a distance of 3½ ft., commencing about five feet one end. The beam contained the heart of the tree, the annular rings being as in the Figure.

Under a load of 35,800 lbs. the beam failed by the tearing apart of the fibres on the tension face.

The maximum skin stress corresponding to this load is 5698 lbs. per square inch.

The co-efficient of elasticity, as determined by an increase in the deflection of .515-ins. between the loads of 2500 and 15,500 lbs., is 1,770,563 lbs.

Table D shows the several readings.

The weight of the beam was 595 lbs. 2 ozs., or 37.76 lbs. per cubic foot on October 3rd, and 583 lbs., or 36.99 lbs. per cubic foot on Nov. 14th, showing a loss of weight in the laboratory at the rate of .0183 lbs. per cubic foot per day.

Table D shews the several readings.

The time occupied by the test was 29 minutes.

Beam XII was tested Nov. 18th, 1893, with the annular rings as in Fig. 17. This beam was cut from a log 28 ins. in diameter, grown probably about 30 feet above the sea level at Port Grey, about eight miles from Vancouver. The tree was felled in August, 1892; it remained in salt water nine months, being alternately wet and dry according to the tide; it was then towed to the mill and cut up.

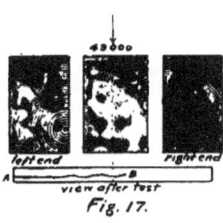

Fig. 17.

The grain was straight and parallel to the axis, and the timber was of good quality corresponding to first quality in the market. It shewed several knots of medium size and a few season cracks. The beam contained the heart of the tree, the annular rings being as in Fig.

Under a load of 49,000 lbs. the beam failed by shearing longitudinally along the season crack AB.

Under this load the maximum skin stress is 7,645 lbs. per sq. in.

The co-efficient of elasticity as determined by an increment in the deflections of .515 ins. between the loads 2,500 lbs. and 15,000 lbs. is 1,678,500 lbs.

Table D shews the several readings.

The time occupied by the test was 37 minutes.

The weight of the beam was 572 lbs., or 35.65 lbs. per cubic foot on Oct. 3rd, and 558 lbs. 4 ozs., or 34.79 lbs. per cubic foot on Nov. 17th showing a loss of weight in the laboratory at the rate of .0191 lbs. per cubic foot per day.

Beam XIII. The history of this beam is the same as that of Beam IX. The beam was tested on Nov. 17th, 1893. The heart of the tree was in one of the faces, the annular rings being as in Fig. 18.

The timber was in good condition and of a quality corresponding to first quality in the market; there were small season cracks along the back of the beam, in the neighbourhood of the neutral plane, and there were also small season cracks along the whole of the front about 3 ins. above the face in compression.

Under a load of 29,300 lbs. this beam failed by the crippling of the fibres on the compression face, commencing at a small knot at the back, Fig. 19.

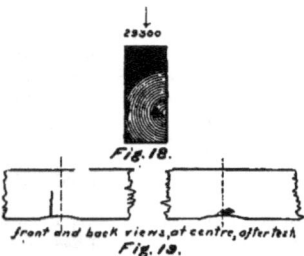

Fig. 18.

front and back views, at centre, after test
Fig. 19.

The maximum skin stress corresponding to this load is 6912 lbs. per square inch.

The co-efficient of elasticity as determined by an increase in the deflection of .805-ins. between the loads 1000-lbs. and 13,000 lbs. is 1,643,193 lbs.

Table E shows the several readings.

The beam weighed 381 lbs. 15 oz., or 34.56 lbs. per cubic foot on Oct. 3rd, and 375 lbs., or 34.13 lbs. per cubic foot on Nov. 15th, showing a loss of weight in the laboratory at the rate of .01 lb. per cubic foot per day.

The time occupied by the test was 45 minutes.

Beam XIV is in reality Beam XIII re-tested, the second test having been made Dec. 2nd, 1893. The beam was replaced in the machine with the crippled side reversed so as to be in tension. The load was then gradually applied until it amounted to 17,600 lbs., when the beam failed on the tension side by the tearing apart of the fibres along the surface at which the crippling took place on the previous test.

The maximum skin stress corresponding to this load is 4082 lbs. per square inch as compared with 6912 lbs. per square inch in the first test. The co-efficient of elasticity, as determined by an increment in the deflection of .51 ins. between the loads of 1,000 lbs. and 8,000 lbs., is 1,513,950 lbs. as compared with 1,643,193 lbs. in the first test.

Table E shews the several readings.

This experiment therefore shows that although the beam may have been crippled by undue pressure, it still retained a large amount of strength as well as elasticity.

Table E gives the several readings.

Beam XV. This beam was tested Nov. 18th, 1893. The timber was excellent in quality, equal to first quality in the market, clear and straight grained and free from knots. Its history is the same as that of Beam XII. The annular rings were oblique as in Fig. 20.

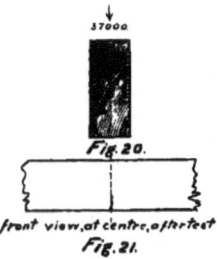

Fig. 20.

front view, at centre, after test
Fig. 21.

Under a load of 37,000 lbs. the beam failed by the crippling of the fibres on the compression face, Fig. 21.

The maximum skin stress corresponding to this load is 8020 lbs. per square inch.

The total compression of the material was .34-in., and the maximum skin compressive stress, taking 1466-in., as the effective depth, is 8189-lbs. per sq. in, the corresponding skin tension tress being 8577 lbs. per in. sq.

Assuming the ordinary law to hold good for the whole of the effective depth, the maximum skin stress would be 8511 lbs. per sq. in.

The co-efficient of elasticity as determined by an increment in the deflection of .755-ins. between the loads, 2000 lbs. and 18,000 lbs., is 1,989,400 lbs.

Table E shews the several readings.

The time occupied by the test was 30 minutes.

The weight of the beam was 445 lbs. 6 ozs., or 39.99 lbs. per cubic foot on Oct. 3rd, and 433 lbs. 13 ozs., or 38.92 lbs. per cubic foot on Nov. 17th, showing a loss of weight in the laboratory at the rate of .0237 lbs. per cubic foot per day.

Beam XVI. This is really Beam XV re-tested, the second test having been made on Dec. 8th, 1893. In the first test the beam had failed by crippling on the compression face; the beam was now reversed, and under a load of 25,580 lbs. it failed by the tearing apart of the fibres on the tension face along the surface at which the crippling had previously taken place. The tensile fracture extended 2 inches below the skin. The jockey weight was now run back until the lever again floated, and the load was gradually increased until it amounted to 32,000 lbs., when the beam fractured a second time on the tension side the fracture extending to a depth of 5 inches below the skin. The first fracture was accompanied by a longitudinal opening (as in Fig.) about 60 inches in extent. A second longitudinal opening, also about 60 inches long, occurred at the second fracture.

The maximum skin stress corresponding to the breaking load of 25,580 lbs. is 5466 lbs. per square inch.

The co-efficient of elasticity, as determined by an increment in the deflection of .54 ins. between the loads of 1,000 lbs. and 11,500 lbs., was 1,825,450 lbs.

Table E gives the several readings.

The weight of the beam was reduced to 428 lbs., or 38.40 lbs. per cubic foot, showing a loss between the test on Nov. 17th, and that on Dec. 8th, at the rate of .02476 lbs. per cubic foot per day.

Beams XVII to XXI were sent to the testing laboratories by the British Columbia Mills Timber & Trading Company through Mr. C. M. Beecher. The whole of these timbers were cut on the coast section of British Columbia. The trees from which Beams XVII, XVIII. XX and XXI were cut, were felled during the summer of 1893, and came from Hartney's Camp, Seymour Creek, while Beam XIX was cut from a tree felled in the spring of 1894, and came from Rowling's Camp, Salmon Arm.

Beam XVII was tested June 24th, 1894. This beam was coarse grained, the grain running very nearly parallel with the axis, and it contained a number of small knots on the compression side. It was cut from the heart of the tree, and was tested with the annular rings as in Fig. 22.

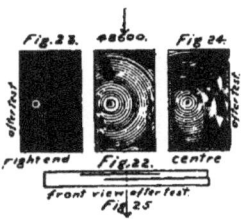

Under a load of 48,600 lbs. it failed by the tearing apart of the fibres on the tension face, the corresponding maximum skin stress, neglecting the compression of the timber, being 4906 lbs. per square inch. The tensile fracture was followed immediately by a longitudinal shear, coincident with the neutral plane at the centre of the beam, and extending for a distance of 8 feet from the end, Fig. 25. The distance between the portions of the beam above and below the plane of shear at the end was 3-10ths of an inch. Figs. 23 and 24 are sections at the end and at the centre showing the nature of the fractures.

The total compression of the material was 1.83 ins., and the maximum skin compressive stress, taking 13.295 ins. as the effective depth, is 5193 lbs. per square inch, the corresponding stress in the tension skin being 6851 lbs. per square inch.

Assuming the ordinary law to hold good for the whole of this effective depth, the maximum skin stress would be 6350 lbs. per square inch.

The co-efficient of elasticity as determined by an increment in the deflection of .335-ins. between the loads 10,000 lbs and 30,000 lbs., is 1,259,600 lbs.

Table F gives the several readings.

The weight of the beam, when shipped from Vancouver about April 21st, was 428 lbs., or 37.21 lbs. per cubic foot ; on reaching the Laboratory on June 9th, the weight was found to be 411 lbs. 10 ozs, or 35.78 lbs. per cubic foot, and on the day of the test, namely, June 24th, the weight was 404 lbs. 8 ozs., or 35.17 lbs. per cubic foot, showing a loss at the rate of .02918 lb. per cubic foot per day between Vancouver and the laboratory, and a loss at the rate of .04067-lb. per cubic foot per day while in the laboratory.

Beam XVIII. This beam was coarse grained, and contained several large and small knots ; it was cut from the heart of the tree. It was tested Sept. 28th, 1894, with the annular rings as in Fig. 26.

The load on the beam was gradually increased to 12,000 lbs. The beam was now gradually relieved from strain until the load had been reduced to 1000 lbs. without showing any set. The load was again gradually increased from 1000 lbs. up to 19,000 lbs., when the beam was again relieved from load and the readings were taken for each difference of 1,000 lbs.

When the load had been reduced to 1000 lbs , the deflection at the centre was observed to be .015-in. as compared with .005-in. in the forward movement, and as soon as the beam was relieved of this 1000 lbs., it returned to its initial condition without showing any set whatever.

The time occupied by the first loading was 10 minutes, by the second loading 12 minutes, and by the relieving from load 8 minutes.

In the final test the load was gradually increased from nil until it amounted to 69,400 lbs., when the beam failed by shearing longitudinally, the shear being immediately followed by the tearing apart of the fibres on the tension face, Figs. 27 28, 29.

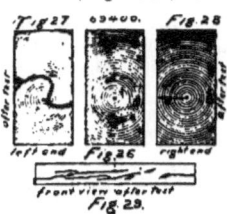

The maximum skin stress corresponding to the breaking load was 5196 lbs. per square inch.

The co-efficient of elasticity, as deduced from an increment in the deflection of 1-10th of an inch between the loads of 2000 lbs. and 12,000 lbs., being 1,329,900 lbs.

Table F gives the several readings.

The weight of the beam at the date of shipment from Vancouver, April 21st, was 512 lbs., or 39.08 lbs. per cubic foot On reaching the laboratory, on June 9th, this weight was 492 lbs. 10 ozs., or 37.60 lbs. per cubic foot, and the weight on Sept. 25th was 466 lbs. 6 ozs., or 35.59 lbs. per cubic foot, showing a loss in weight between Vancouver and the laboratory at the rate of .0302-lb. per cubic foot per day, and a loss of weight in the laboratory at the rate of .0181-lb. per cubic foot per day.

Beam XIX. This beam was of exceptionally good quality, with clear close grain and no knots. It was tested Oct. 2nd, 1894, with the annular rings nearly vertical, as in Fig. 30.

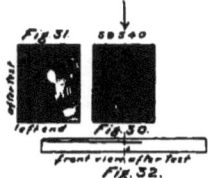

The load on the beam was gradually increased up to 16,000 lbs., when it was gradually relieved from load, the readings being taken for each diminution of 4000 lbs. The corresponding readings are indicated in Table F.

When it was completely relieved from load, the scales showed readings of .005-in at the centre, .001-in and .003 in at the ends. These readings were probably due to inequalities in the timber or a possible sliding of the scales, as the beam showed no evident sign of set.

The load was again immediately increased gradually from nil until it amounted to 59,540 lbs., when the beam failed by longitudinal shear, followed by the splintering of the upper edges on the tension side, Figs. 31, 32. Fracture was also indicated by the crippling of the fibres on the compression side taking place between 58,000 and 59,540 lbs.

The distance between the portions of the beam above and below the plane of shear at the end was .36-in. as in the figure.

The maximum skin stress corresponding to the breaking load is 9043 lbs. per square inch.

The co-efficient of elasticity, as deduced by an increase in the deflection of .3-in. between the loads of 2000-lbs. and 16,000 lbs., is 1,934,600 lbs.

Table F shows the several readings.

The time occupied by the first loading was 10½ mins., by the relieving from the load 6¾ mins., and by the second loading from nil to the max., 15¼ mins.

The weight of this beam on April 21st, the date of its shipment from Vancouver, was 410 lbs., or 44.99 lbs. per cubic foot. On reaching the laboratory the weight was 392 lbs. 8 ozs., or 43.07 lbs. per cubic foot, and the weight on Oct. 2nd, the date of the test, was 375 lbs. 10 ozs., or 41.22 lbs. per cubic foot, showing a loss of weight at the rate of .0392-lb. per cubic foot per day between Vancouver and the laboratory, and a loss at the rate of .0161-lb. per cubic foot per day while in the laboratory.

Beam XX. This beam was cut from the heart of the tree, and was tested Nov. 3rd., 1894, with the annular rings as in Fig. 33.

It was coarse grained, the grain being very nearly parallel with the axis, and contained a number of knots.

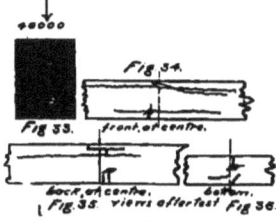

The load was gradually increased until it amounted 12,000 lbs., and at this point the beam was gradually relieved from load, readings being taken for every diminution of 2000 lbs. When the load had been reduced to 500 lbs., the reading at the centre was .001-in., probably due to a movement of the scale. The load was again gradually increased until it amounted to 40,000 lbs., when the beam failed by the crippling of the fibres on the compression side in the neighbourhood of a small knot 1¼ in. above the compression face, Figs. 34, 35, 36. The crippling extended about 4 ins. above this face. The load was still gradually increased until it amounted to 49,600 lbs., when the beam again failed by the tearing apart of the fibres on the tension face.

The maximum skin stress corresponding to the load of 40,000 lbs., and disregarding the compression of the timber, is 6559 lbs., and the skin stress corresponding to the load of 49,600 lbs., is 8127 lbs. per square inch.

The total compression of the timber was .345-ins., so that taking the effective depth under this load to be 11.655 ins., the maximum skin compressive stress would be 6710 lbs. per square inch, the corresponding skin tension stress being 7125 lbs. per square inch.

Assuming the ordinary law to hold good for the whole of the effective depth, the maximum skin stress would be 6936 lbs per square inch.

The co-efficient of elasticity, as deduced from a change in the deflection of .22-in. between the loads 4000 lbs. and 12,000 lbs., both forwards and while being relieved from load in the first reading, and also during the second loading, is 1,571,150 lbs.

Table G shows the several readings.

The weight of this beam when shipped from Vancouver, April 21st, was 349 lbs, or 41.16 lbs. per cubic foot ; when delivered at the laboratory on June 9th, it weighed 329 lbs., or 36.70 lbs. per cubic foot, and on Nov. 3rd it weighed 311 lbs. 6½ ozs., or 34.92 lbs. per cubic foot, showing a loss of weight between Vancouver and the laboratory at the rate of .091-lb. per cubic foot per day, and a loss while in the laboratory at the rate of .0121-lb. per cubic foot per day.

The time occupied by the test was 26 mins.

Beam XXI. This beam was tested Nov. 3rd, 1894, with the annular rings as in Fig. 37.

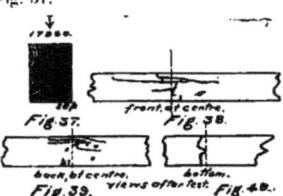

Fig. 37. Fig. 38.
Fig. 39. views after test. Fig. 40.

The load upon the beam was gradually increased until it amounted to 6000 lbs., when it was gradually relieved of load, at the rate of 1000 lbs. for each observation, and the beam returned to its initial condition without showing any sign of set. The load was again gradually increased until it amounted to 17,960 lbs., when a sharp fracture took place by the tearing apart of the fibres on the tension side, and this was accompanied by a simultaneous crippling of the fibres on the compression side, Figs. 38, 39, 40.

The maximum skin stress corresponding to the load of 17,960 lbs. is 7787 lbs. per square inch.

The total compression of the timber at the centre was .16-in., so that taking the effective depth at the centre to be 8.82 ins., the maximum skin compressive stress at the point of fracture is 7901 lbs. per square inch, the corresponding skin tensile stress being 8221 lbs. per sq. in.

Assuming the ordinary law to hold good for the whole of the effective depth, the max. skin stress would be 8100 lbs. per sq. in.

The co-efficient of elasticity, as deduced by a change in the deflection of .48-in. between the loads of 1000-lbs. and 6000 lbs., during the first loading, and while being relieved of load, is 1,588,400 lbs.

Table G shows the several readings.

The weight of this beam when shipped from Vancouver, April 21st, was 164 lbs., or 38.86 lbs. per cubic foot ; when received at the laboratory on June 9th, the weight was 151 lbs. 4 ozs., or 33.02 lbs. per cubic foot, and on Nov. 13th, the date of test, the weight was 139 lbs. 10½ ozs., or 30.83 lbs. per cubic foot, showing a loss of weight between Vancouver and the laboratory at the rate of .1192 lbs. per cubic foot per day, and a loss of weight while in the laboratory at the rate of .0149 lbs. per cubic. foot per day.

The time occupied by the test was 18½ mins.

OLD DOUGLAS FIR.

Beams XXII–XXV were sent to the Laboratory by Mr. P. A. Peterson, Chief Engineer of the Canadian Pacific Railway.

These beams were four old stringers taken from trestles numbered 428, 35, 316 and 789.

Trestle 428 is about half way between Cisco Cantilever Bridge and Lytton. It was erected in the early summer of 1884, and the timbers had consequently been in position for nine years. It is in a dry country, with very little rainfall, and subject to a hot sun in summer. The stringer from this structure was cut out of a log probably grown on a flat about three miles west of Hope, where most of the trees were windshaken.

Trestle No. 35 is about one mile west of Port Moody, and was built in the early spring of 1887, so that the stringer was in position for a period of 6½ years in a place subject to the heaviest rainfall in the province. The stringer was cut from a log most probably grown at Point Grey, about eight miles from Vancouver.

Trestle No. 316 is two miles east of Spuzzum. The stringer from this trestle was cut from a log grown on a bench near Spuzzum about 500 feet above the sea-level. It was prepared and framed in 1881, and erected in 1882, so that it was eleven years in position in a district with a climate similar to that of Nova Scotia. As the railway here runs north and south, the sun had not the same effect upon the stringers as on other parts of the line.

Trestle No. 789 is on Kamloops Lake, six miles east of Savona, and was erected in the spring of 1885, so that the timbers had been in service for a period of eight years. The neighbourhood is dry, but the trestle, being situated under a high bluff, is protected from the afternoon sun. The stringer from this structure was cut out of a log probably grown about three miles west of Hope, at the same place as the timbers used in structure No. 428.

Beam XXII from Trestle 428, was tested Nov. 25th, 1893, with the annular rings as in Fig. 41.

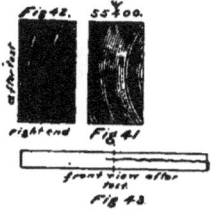

There were two vertical 1-in. bolt holes in the timber,—one near the centre and one at the end. There were also several season cracks in the timber, one being somewhat large.

The load upon the beam, was gradually increased until it amounted to 55,400 lbs., when the beam failed by a longitudinal shear, as in Figs. 42, 43.

The distance between the portions of the beam above and below the plane of shear at the end was ⅜ths of an inch.

The maximum skin stress corresponding to the breaking load is 7086 lbs. per square inch.

The total compression of the timber at the centre was .63 in., so that taking the effective depth at 15.0575 ins., the maximum skin compressive stress is 7264 lbs. per square inch, the corresponding tensile skin stress being 7898 lbs. per square inch.

Assuming the usual law to hold good for the whole of the effective depth, the maximum skin stress would be 7,382 lbs. per square inch.

The co-efficient of elasticity, as deduced by an increase in the deflection of .39 in. between the loads of 2,000 lbs. and 20,000 lbs., is 1,639,500 lbs., while it is 1,691,620 lbs. for an increment in the deflection of .42 in. between the loads 2,000 lbs. and 22,000 lbs.

Table II gives the readings under the several loads.

The weight of the beam on the day of test was 33.75 lbs. per cubic foot, and the total weight on Oct. 3rd was 438 lbs. 7 ozs.

Beam XXIII from Trestle No. 789 was tested Nov. 28th, 1893, with the annular rings as in Fig. 44, and showing the heart in one of the faces.

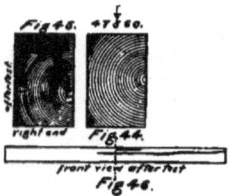

Fig. 46.

The load upon the beam was gradually increased until it amounted to 47,560 lbs., when the beam failed by the tearing apart of the fibres on the tension face, which was immediately followed by a longitudinal shear, as in Figs 45, 46.

The maximum skin stress corresponding to the load of 47,560 lbs. is 7339 lbs.

The co-efficient of elasticity, as deduced from an increment of .06 in. in the deflection between the loads of 2,000 lbs. and 22,000 lbs., is 1,878,950 lbs.

Table I shows the readings under the various loads.

The total weight of the beam on Oct. 3rd was 654 lbs. 12 ozs., or 38.95 lbs. per cubic foot; the total weight on Nov. 28th, the date of test, was 549 lbs. 8½ ozs., or 38.59 lbs. per cubic foot, showing a loss of weight in the laboratory at the rate of .00643 lbs. per cubic foot per day. Estimating the weight of this beam from a solid block cut out of the beam, it was found to be 39.13 lbs. per cubic foot, or .54 lb. per cubic foot heavier than the weight deduced from the total weight of the whole beam.

Beam XXIV from Trestle No. 35. This beam was tested Nov. 25th, 1893, with the annular rings as in Fig. 47. It contained two vertical ⅜-in. bolt holes about half way between the centre and ends, and a few knots of average size appeared on the face. It also contained several season cracks.

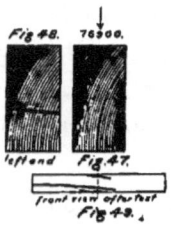

Fig. 49.

The initial load, including the weight of the beam, was 5,000 lbs., and the load was gradually increased up to 41,000 lbs., when the material at one end of the beam was crushed in. The ends of the beam were found to be very much the worse for wear and in a rotten condition. Releasing the beam from load the ends were sawn off and the beam was replaced at 9-ft. centres, when the load was gradually increased until it amounted to 76,900 lbs. Under this load the beam failed by longitudinal shear, which was accompanied by a certain amount of crippling of the fibres on the compression side of the centre, as in Figs. 48, 49.

The maximum skin stress corresponding to the breaking load of 76,900 lbs. was 6135 lbs. per square inch.

The total compression under a load of 41,000 lbs. at the centre was 1.7 in., and taking the effective depth of the beam to be 14.5 ins., the corresponding maximum skin compressive stress is 6495 lbs. per square inch, the corresponding skin tensile stress being 8221 lbs. per square inch.

Assuming the ordinary law to hold good for the whole of the effective depth, the maximum skin stress would be 7662 lbs. per square inch.

The co-efficient of elasticity, as determined by an increase in the deflection of .16-in. between the loads of 11,000 and 22,000 lbs., is 1,199,741 lbs.; as determined by an increment of the deflection of .33-in. between the loads 10,000 lbs. and 32,000 lbs, it is 1,163,354 lbs.; and as deduced from an increment in the deflection of .29 in., the mean between .285-in. and .295-in., the increments between the loads of 5,000 and 25,000 lbs. and 10,000 and 30,000 lbs. respectively, it is 1,203,500 lbs.

Table II shows the several readings.

The total weight of the beam on Nov. 25th, the date of test, was 331 lbs. 9 ozs., or 32.8 lbs. per cubic foot. After cutting off the ends, the weight of a length of 9 feet was 262 lbs. 5 ozs., or 33.4 lbs. per cubic foot. The total weight of the beam on October 3rd was 339 lbs. 9 oz.

Beam XXV from Trestle 316. This beam was tested Nov. 28th, 1893, with the annular rings as in Fig. 50, and showing the heart on one of the faces.

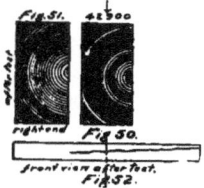

It contained one vertical bolt hole, several knots, and many season cracks. The grain was straight.

The load upon the beam was gradually increased until it amounted to 42,900 lbs., when a large splinter broke off on the tension fee, and the beam failed by longitudinal shear, as in Figs. 51, 52.

The maximum skin stress corresponding to this breaking load is 4613 lbs. per square inch.

The co-efficient of elasticity, as determined by an increment in the deflection of .335-in. between the loads of 4,000 lbs. and 20,000 lbs., is 949,720 lbs.

Table I shows the readings for the several loads.

The total weight of the beam on October 3rd was 422 lbs., or 34.44 lbs. per cubic foot, and on Nov. 28th, the date of test, the weight was 406 lbs., or 33.11 lbs. per cubic foot, showing a loss of weight in the Laboratory at the rate of .237-lbs. per cubic foot per day.

The time occupied by the test was 30 minutes.

The following Table gives a summary of the results obtained for Douglas Fir:—

BEAM.	Dimensions in inches.			Weight in lbs. per cubic foot at date of test.	Maximum skin stress in lbs. per sq. in.	Co-efficient of elasticity in lbs.
	l	*d*	*b*			
NEW TIMBER. SPECIALLY SELECTED.						
III.	66 ×	5.375 ×	4.125		10,441	2,178,100
XIX.	138 ×	12.1 ×	9.1	41.22	9,643	1,934,500
VII.	89 ×	6 ×	5.8125	39.92	8,712	2.044,115
XV.	198 ×	15 ×	6.125	38.92	8,020	1,989,400
NEW TIMBER, FIRST QUALITY.						
	l	*d*	*b*			
X	198 ×	14.875 ×	6	37.50	4,027	1,629,616
XI	204 ×	14.875 ×	8.6875	36.99	5,698	1,770,563
IX	204 ×	14.875 ×	9	35.76	7,691	1,764,989
VIII	69 ×	5.125 ×	5.5	35.74	8,382	1,584,692
XVIII	138 ×	17.8 ×	8.76	35.59	5,196	1,329,900
XVII	138 ×	15.125 ×	9.	35.17	4,907	1,259,600
XX	138 ×	12. ×	8.88	34.92	6,559	1,571,150
XII	204 ×	14.875 ×	8.8125	34.79	7,645	1,678,300
XIII	204 ×	14.75 ×	66	34.13	6,912	1,643,193
XXI	138 ×	8.98 ×	5.95	30.83	7,784	1,588,100
VI	69 ×	6.125 ×	6	30.23	7,116	1,489,215
I	96 ×	12.125 ×	9.		4,897	1,138,500
II	66 ×	12.125 ×	5.625		4,378	1,140,900
V	69 ×	9.125 ×	5	29.18	5,869	946,270
IV	69 ×	9.125 ×	5.	28.27	4,156	926,500
OLD TIMBER.						
	l	*d*	*b*			
XXIII	186 ×	14.85 ×	8.78	39.59	7,339	1,878,950
XXII	162 ×	15.6875 ×	7.75	33.76	7,086	1,665,560
XXV	144 ×	15.65 ×	8.2	33.11	4,613	949,720
XXIV	132 ×	16.2 ×	7.75	32.8	6,135	1,201,020

The following data may be adopted in practice:—

In the case of specially selected timber, free from knots, with sound clear and straight grain, and cut out of the log at a distance from the heart:

Average weight in lbs. per cubic foot = 40.
Average co-efficient of elasticity in lbs. per sq. in. = 2,000,000.
Average maximum skin stress in lbs. per square inch = 9000.
Safe working skin stress in lbs. per square inch = 3000-lbs.

In the case of first quality timber, such as is ordinarily found in the market:

Average weight in lbs. per cubic foot = 34.
Average co-efficient of elasticity in lbs. per square inch = 1,430,000.
Average maximum skin stress in lbs. per square inch = 6000.
Safe working skin stress in lbs. per square inch = 2000.

In specifying these data it will be observed that 3 is adopted as the factor of safety. Upon this hypothesis the factor of safety for the stick giving the minimum skin stress in more than 2, and this, in the opinion of the author, is an ample factor for a material which experience and all experiments show, may be strained without danger very nearly up to the point of fracture.

Further, the results obtained in the experiments with the old stringers show that the strength of the timber had been retained to a very large extent, and that the rotting had not extended to such a depth below the skin as to sensibly affect the efficiency of the sticks, which still possessed ample strength for the work they were designed to do.

Thus in Beam XXII a diminution in the skin stress of 1058 lbs. per square inch, which is equivalent to a diminution in the effective depth of $\frac{15.6875 \times 1058}{2 \times 7086}$ = 1.076-ins. would still leave 6000 lbs. per square inch as the skin stress. Thus if the rotting had extended to depth of 1.176 ins., the factor of safety would still remain 3.

If 2 is adopted as the factor of safety, and, in the opinion of the author, 2 is an ample factor for the great majority of cases, the rotting might extend without danger to a depth of 3.398 ins.

In the case of Beam XXV, which is the old stringer giving the least co-efficient of strength, namely, 4613 lbs. per square inch, taking 2 as the factor of safety, the effective depth might be diminished by an amount of $\frac{17.65 \times 613}{2 \times 4613} = 1.04$ ins. and rot might safely extend to this depth.

Again, it will be observed that the skin stress and the elasticity are subject to a wide variation. This variation is due to many causes, of which the most important are the presence of knots, obliquity of grain, and, more than all, the locality in which the timber was grown, the original position of the stick in the log from which it was cut, and the proportion of hard to soft fibre, or of the summer to the spring growth. The tensile shearing and compressive experiments upon specimens cut out of different parts of the same log all shew that the timber near the heart possesses much less strength and stiffness than the timber at a distance from the heart.

The accompanying photograph is given to show the variation of

BEAM XIII   BEAM IX

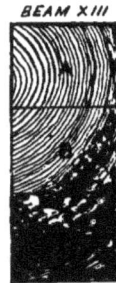 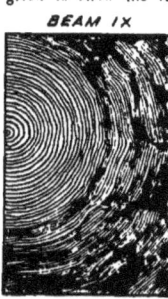

thickness in the growth rings from the heart outwards, and a careful study of the results obtained up to date would seem to indicate that the best classification defining the strength of the timber would be found by dividing the section of a log into three parts by means of two circles, with the heart as the centre, and by designating the central portion as third quality, the portion between the two circles as second quality, and the outermost portion as first quality.

A most interesting paper on the structural characteristics of Douglas Fir from a botanical standpoint was read by Professor Penhallow, F.R.S.C., at the meeting of the Royal Society of Canada in Ottawa, in 1894, in connection with a paper by the author on the strength of the timber.

RED PINE.

Beams XXVI to XXXIII were sent to the laboratory by Messrs. McLachlin Bros., of Arnprior.

These beams were not specially selected, but were the ordinary scantlings in the market. They were cut from logs felled in February or March, 1893, in the neighbourhood of the Bonnechère River, Nipissing District, County Renfrew. The logs remained in the water from April until October, when they were sent to the mill, where they were sawn up and piled.

Beam XXVI. This beam was cut from the heart of the tree, and was tested March 13th, 1894, with the annular rings, as in Fig. 53.

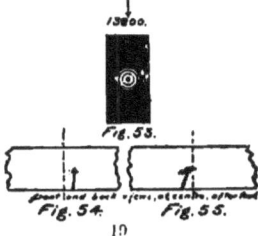

The load upon the beam was gradually increased until it amounted to 13,800 lbs., when the beam failed by the crippling of the fibres on the compression face, Figs. 54, 55. The load was still further increased until complete fracture took place by the tearing apart of the fibres on the tension face under a load of 17,170 lbs. The crippling was in line with a knot running through the timber from back to front, as in the Figure.

The maximum skin stress corresponding to the load of 13,800 lbs. is 3937 lbs. per square inch.

The total compression of the timber at the centre was .2-in., so that, taking the effective depth as 13.05, the maximum skin compressive stress would be 3994 lbs. per sq. in., the corresponding skin tensile stress being 4119 lbs. per square inch.

Assuming the ordinary law to hold good for the whole of the effective depth, the maximum skin stress would be 4059 lbs. per square inch.

The co-efficient of elasticity, as determined by an increment in the deflection of .885-in. between the loads 1,000 and 8,000 lbs., is 1,235,600 lbs., and as determined by an increment in the deflection of .5-in. between the loads 2,000 and 6,000 lbs., is 1,248,990 lbs.

Table K shows the several readings.

The weight of this beam, on March 10th, was 392 lbs. 2 ozs., or 37.56 lbs. per cubic foot, and on March 13th it was 379 lbs. 4 ozs., or 36.39 lbs. per cubic foot, showing a loss of weight in the laboratory at the rate of .39-lb. per cubic foot per day.

Beam XXVII was tested April 5th, 1894, with the annular rings as in Fig. 56. The beam was cut from the heart of the tree, and the darkened portion in the Figure, was sapwood.

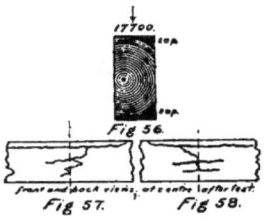

The load upon the beam, was gradually increased until it amounted to 17,700 lbs., when the beam failed by the tearing apart of the fibres on the tension face, Figs. 57, 58, at a resin pocket, the fracture showing a fine resinous surface.

The maximum skin stress corresponding to the breaking load is 5219 lbs. per square inch.

The total compression of the timber at the centre was .34-in., so that taking 12.785 ins. as the effective depth, the maximum skin compressive stress would be 5111 lbs. per square inch, the corresponding skin tensile stress being 5707 lbs. per square inch.

Assuming the ordinary law to hold good for the whole of the effective depth, the maximum skin stress would be 5501 lbs. per square inch.

The co-efficient of elasticity, as deduced from an increment in the deflection of 7-in. between the loads 1500 lbs. and 7500 lbs., is 1,418,500 lbs.

Table K gives the several readings.

The total weight of the beam on March 10th was 46 lbs. 12 ozs., or 44.51 lbs. per cubic foot ; the total weight on April 5th, the date of test, was 397 lbs. 4 ozs., or 36.50 lbs. per cubic foot, showing a loss of weight while in the laboratory, at the rate of .192-lbs. per cubic foot per day.

Beam XXVIII. This beam was cut from the heart of the tree, and was tested April 20th, 1894, with the annular rings as shown in Fig 59.

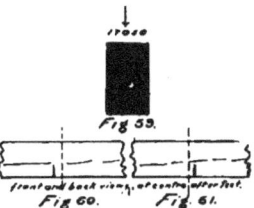

Fig 59.

Fig 60.   Fig. 61.

The load upon the beam was gradually increased until it amounted to 17,050 lbs., when the beam failed by the crippling of the fibres on the compression face, Figs. 60, 61. The load was still increased until under 19,140 lbs. the beam again failed by the tearing apart of the fibres on the tension face.

The maximum skin stress corresponding to the load under which crippling took place is 6752 lbs. per square inch.

The total compression of the beam under a load of 17,050 lbs. was .24 in., so that taking the effective depth to be 11.01 ins., the corresponding maximum skin compressive stress would be 6886 lbs. per square inch, the corresponding skin tensile stress being 7193 lbs. per square inch.

Assuming the usual law to hold good for the whole of the effective depth, the maximum skin stress would be 7050 lbs. per square inch.

The co-efficient of elasticity, as determined by an increase in the deflection of 1,435 in between the loads of 2000 and 12,000 lbs., is 1,786,000 lbs.; it is 1,858,400 lbs., as determined by an increment in the deflection of .84-in. between the loads 3500 and 9500 lbs., and is 1,681,100 lbs., as determined by an increment in the deflection of 1.135 in. between the loads of 2000 and 10,000 lbs.

Table K shows the several readings.

The test occupied 26 minutes.

The weight of the beam on March 10th was 379 lbs. 10 ozs., or 44.20 lbs. per cubic foot; upon April 20th, the date of test, the weight was 322 lbs. 8 ozs., or 37.55 lbs. per cub. ft., showing a loss of weight at the rate of .1622-lb. per cubic foot per day.

Beam XXIX. This beam was cut from the heart of the tree, and was tested March 13th, 1894, with the annular rings as in Fig. 62

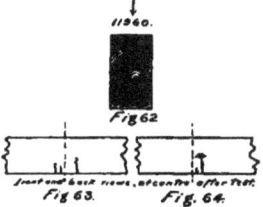

Fig 62.

Fig 63.   Fig. 64.

The load upon the beam, was gradually increased until it amounted to 11,960 lbs., when the beam failed by the crippling of the fibres on the compression face, Figs. 63, 64. The load was still further gradually increased to 12,460 lbs., when the beam was completely fractured by the tearing apart of the fibres on the tension face.

The maximum skin stress corresponding to the breaking load of 11,960 lbs. is 4818 lbs. per square inch.

The total compression of the timber at the centre was .15-in., so that taking 11.1-in. as the effective depth, the maximum skin compressive stress would be 4883 lbs. per square inch, the corresponding skin tensile stress being 5016 lbs. per square inch.

Assuming the usual law to hold good for the whole of the effective depth, the maximum skin stress would be 4949 lbs. per square inch.

The co-efficient of elasticity, as determined from an increment of .86-in. in the deflection between the loads of 1000 and 5000 lbs., is

1,210,100 lbs. The co-efficient of elasticity, as deduced from an increment of 1.315-in in the deflection between the loads of 1000 lbs. and 7000 lbs., is 1,187,000 lbs.

Table L shews the several readings.

The test occupied 27 minutes.

The total weight of the beam was 290 lbs., or 32.89 lbs. per cubic foot on March 10th, and 282 lbs. 6 ozs., or 32.03 lbs. per cubic foot on March 13th, showing a loss of weight in the laboratory at the rate of .2866-lb. per cubic foot per day.

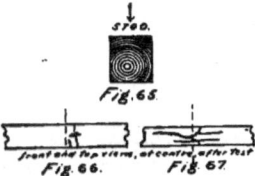

Fig. 65.

Fig. 66. Fig. 67.

Beam XXX. This beam was tested May 3rd, 1894, with the annular rings, as in Fig. 65. When the beam was placed in position, it showed an upward camber of 24 ins.

The load upon the beam was gradually increased until it amounted to 5700 lbs., when the beam failed by the crippling of the fibres on the compression face, Fig. 66, the crippling extending 2½ ins. upwards from the skin. The load was still increased, and when it amounted to 6380 lbs., the beam broke right across the tension face about 2½ inches from the middle of the beam, and vertically above the second line of crippling on the compression side, Fig. 67.

The maximum skin stress corresponding to the breaking load of 5700 lbs. is 4634 lbs. per square inch, and the maximum skin stress corresponding to the load of 6580 lbs. is 5340 lbs. per square inch.

The co-efficient of elasticity is 1,322,000 lbs., as determined by an increment in the deflection of 1.69-in. between the loads of 1000 and 5000 lbs.; it is 1,329,900 lbs., as deduced from an increment in the deflection of .84-in. between the loads of 2000 and 4000 lbs.

Table L shows the several readings.

The weight of this beam on May 4th, the day after the test, was 150 lbs. 11 ozs., or 30.96 lbs. per cubic foot.

Beam XXXI. This beam was tested May 4th, 1894. It was cut from the heart of the tree, and the annular rings were situated as in Fig. 68. Season cracks ran intermittently from end to end of the beam

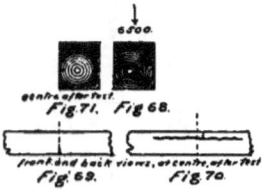

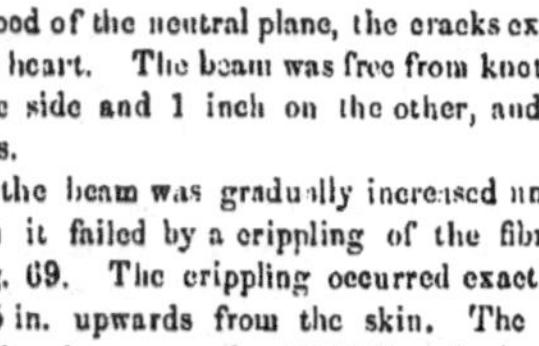

Fig. 71. Fig. 68.

Fig. 69. Fig. 70.

in the neighbourhood of the neutral plane, the cracks extending radially outwards from the heart. The beam was free from knots for a distance of 7 inches on one side and 1 inch on the other, and the grain ran parallel to the axis.

The load upon the beam was gradually increased until it amounted to 6500 lbs., when it failed by a crippling of the fibres on the compression face, Fig. 69. The crippling occurred exactly at the centre and extended 1.5 in. upwards from the skin. The load was then continued, and, when it amounted to 7000 lbs., the beam failed by the tearing apart of the fibres on the tension face, Figs. 70, 71, and a line of crippling on the compression side timber opened upwards for a distance of about 2 ins. or 3½ ins. The fracture on the tension side took place about 5½ ins. from the centre, and the timber opened along the annular rings for a distance of 24 ins. on each side of the centre as in the figure.

The maximum skin stress corresponding to the breaking load of 6500 lbs. is 5442 lbs. per square inch.

The co-efficient of elasticity, as deduced from an increment in the deflection of 1.085 ins. between the loads of 2000-lbs. and 5000 lbs., was 1,618,900 lbs.

Table L shews the several readings.

This beam when first placed in position, also had a camber of .35-ins. in a central length of 14 ft. 6 ins.

The weight of the beam on May 4th, the date of test, was 165 lbs. 6 ozs., or 34.97 lbs. per cubic foot.

Beams XXXII to XXXV might perhaps more properly be designated 3 ins. planks.

Beam (Plank) XXXII was tested May 7th, 1894. The heart was in one of the faces, and the annular rings were situated as in Fig. 72.

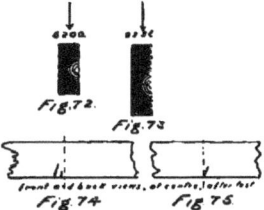

Fig. 72. Fig. 73.

Fig. 74. Fig. 75.

The load upon the beam gradually increased until it amounted to 5200 lbs., when it failed by a crippling of the fibres on the compression side. The crippling occurred about 1½ ins. away from the centre of the beam and extended upwards about 1.5 ins. The load was still increased, and when it amounted to 5860 lbs. the beam again failed by the tearing apart of the fibres on the tension side. A line of crippling also extended upwards a further distance of about 2 ins., or about 3½ ins. from the skin.

The maximum skin stress corresponding to the breaking load of 5200 lbs. is 6928 lbs. per square inch.

The co-efficient of elasticity, as deduced from an increment in the deflection of 1.67-ins. between the loads 1000-lbs. and 4000 lbs., is 1,575,200 lbs. per square inch.

Table L shews the several readings.

The weight of this beam on May 7th, the date of test, was 102 lbs., or 31.56 lbs. per cubic foot.

Beam (Plank) XXXIII was tested May 7th, 1894, with the annular rings as shown in Fig. 73.

The load upon the beam was gradually increased to 9250 lbs., when failure took place by the crippling of the fibres on the compression side, Figs. 74, 75. There were two lines of crippling on the front and one at the middle of the beam at the back. The crippling at the back probably occurred first, as the folding of the timber extends across the section of the beam along the central line at the lower edge, but not up to the point where the failure due to compression was apparently the greatest. In the neighbourhood of the crippling in front, the timber was clear, and the grain ran straight and parallel with the axis; at the back there were three knots, which were primarily the cause of the crippling.

When the load on the beam had been increased to 9900 lbs., fracture occurred on the tension side.

The maximum skin stress corresponding to the breaking load of 9250 lbs. is 6554 lbs. per sq. in.

The co-efficient of elasticity, as determined by an increment in the deflection of .76 in. between the loads 2600 and 6200 lbs., is 1,618,000 lbs.

Table M shews the several readings.

The weight of the beam on May 7th, date of test, was 128 lbs. 8 ozs., or 31.87 lbs. per cubic foot.

Beam (Plank) XXXIV. This beam was tested May 8th, 1894, with the annular rings as in Fig. 76.

The load upon the beam was gradually increased until it amounted to 5600 lbs., when the fibres on the compression face crippled to a small extent. On still further increasing the load, the fibres on the compression face were completely crippled, Figs. 77, 78, and fracture also simultaneously occurred on the tension side when the load amounted to 8400 lbs.

The grain of this beam was straight and parallel with the axis, and the timber was apparently free from knots for a distance of about 24 inches on each side of the centre.

The maximum skin stress corresponding to the breaking load of 5600 lbs. is 5079 lbs. per square inch, and the skin stress corresponding to the load of 8400 lbs., which caused the fracture on the tension side, is 7597 lbs. per square inch.

The co-efficient of elasticity, as deduced from an increment in the deflection of 1.14 ins. between the loads of 500 and 5600 lbs., was 1,784,800 lbs.

Table M shows the several readings.

The weight of the beam on May 8th, date of test, was 96 lbs. 2 ozs., or 36.59 lbs. per cubic foot.

Beam (Plank) XXXV was tested May 8th, 1894, with the annular rings as in Fig. 79. The heart of the tree was very nearly coincident with the axis of the beam, and the grain ran in the same direction. Season cracks occurred intermittently throughout the beam.

The load upon the beam was gradually increased until it amounted to 7600 lbs., when the beam failed by the crippling of the fibres on the compression face, Fig. 80. The load was still increased, and well defined crippling occurred when it amounted to 10,050 lbs. When the load had reached 13,700 lbs. the beam failed by the tearing apart of the fibres on the tension face, Fig. 80.

The maximum skin stress corresponding to the breaking load of 7600 lbs. is 4339 lbs. per square inch.

The co-efficient of elasticity, as determined by an increment in the deflection of .92-in. between the loads of 500 and 7600 lbs., is 1,589,250 lbs., and as determined by an increment in the deflection of .025-in. for the corresponding increase of 200 lbs. it is 1,642,900 lbs.

Table M shows the several readings.

The weight of the beam on May 8th, date of test, was 128 lbs. 12 ozs. or 37.69-lbs. per cubic foot.

The following Table gives a summary of the results obtained, for Red Pine:—

Beam.	Dimensions in inches.			Weight in lbs. per cubic foot at date of test.	Maximum skin stress in lbs. per sq. inch.	Co-efficient of elasticity in lbs.
	l	*d*	*b*			
		NEW TIMBER.				
XXXV.	156 ×	11.15 ×	3.325	37.69	4,339	1,616,075
XXVIII.	210 ×	11.25 ×	6.34375	37.55	6,752	1,802,633
XXXIV.	156 ×	9.125 ×	3.125	36.59	5,079	1,784,800
XXVII.	210 ×	13.125 ×	6.1875	36.50	5,219	1,418,500
XXVI.	210 ×	13.25 ×	6.375	36.39	3,937	1,241,950
XXXI.	174 ×	7.125 ×	6.21875	34.97	5,442	1,614,900
XXIX.	210 ×	11.25 ×	6.25	32.03	4,818	1,198,550
XXXIII.	180 ×	11.125 ×	3.1	31.87	6,554	1,618,000
XXXII.	180 ×	8.125 ×	3.1	31.56	6,928	1,575,200
XXX.	174 ×	7.25 ×	6.1875	30.96	4,634	1,325,950

Hence,

The average weight in lbs. per cubic foot = 34.61.
  " co-efficient of elasticity in lbs. per sq. in. = 1,520,036.
  " maximum skin stress " " = 5370.

If, however, the plank results are omitted,
The average weight in lbs. per cubic foot = 34.78.
  " co-efficient of elasticity in lbs. per sq. in. = 1,434,747.
  " maximum skin stress " " = 5137.

In general, the following data may be adopted in practice:—
The average weight in lbs. per cubic foot = 34.6.
  " co-efficient of elasticity in lbs. per sq. in. = 1,430,000.
  " maximum skin stress " " = 5,100.
  " safe working skin stress " " = 1,700,
3 being a factor of safety.

In the accounts of the several beams it will be observed that the failures are almost invariably due to the crippling of the material on the side in compression, indicating that the tensile strength of the timber exceeds its compressive strength, and this was subsequently verified by the direct tension and compression experiments.

WHITE PINE.

Beams XXXVI and XXXVII are two pieces cut out of one large piece of square pine, made and taken out in the Gatineau Valley, Ottawa County. The timber was brought down via the Gatineau and Ottawa Rivers to Montreal, and remained in the water until late in the fall of 1892, when it was piled on the land for winter sawing.

This timber was purchased from Messrs. J. & B. Grier.

Beam XXXVI was tested February 16th, 1893, with the annular rings as in Fig. 81.

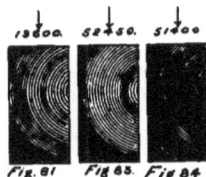

Fig. 81. Fig. 83. Fig. 84.

The load upon the beam was gradually increased until it amounted to 19,000 lbs., when it failed by the tearing apart of the fibres on the tension side.

The maximum skin stress corresponding to this load is 2993 lbs. per square inch.

The co-efficient of elasticity, as determined by an increment in the deflection of 1.12 ins. between the loads of 5000 and 10,000 lbs., is 503,440 lbs.; as deduced from an increment in the deflection of .84-in. between the loads of 5000 and 12,500 lbs., is 463,768 lbs., and as deduced from an increment in the deflection of 2.13 ins. between the loads of 5000 and 15,000 lbs., is 534,169 lbs.

Table N shows the several readings.

The weight of this beam per cubic foot on Feb. 16th was 37.25 lbs.; and on March 14th, 34.78 lbs., showing a loss of weight at the rate of .095-lb. per cubic foot per day.

Beam XXXVII was tested on February 24th, 1893, with the annular rings as in Fig. 82.

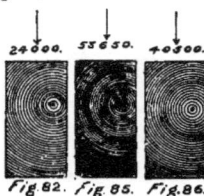

Fig. 82.   Fig. 85.   Fig. 86.

The load was gradually increased until it amounted to 24,000 lbs., when the beam failed by the tearing apart of the fibres on the tension face.

The maximum skin stress corresponding to this load is 3555 lbs. per square inch.

Beams XXXVIII and XXXIX were the two ends of Beam XXXVI which was tested February 16th, 1893, the central portion containing the fracture having been cut out.

Beam XXXVIII was tested on March 14th, with the annular rings as in Fig. 83.

The load on the beam was gradually increased until it amounted to 52,450 lbs., when it failed by the tearing apart of the fibres on the tension side.

The maximum skin stress corresponding to this load is 3075 lbs. per square inch.

The co-efficient of elasticity, as determined by an increment in the deflection of .37-in. between the loads of 10,000 and 25,000 lbs., is 622,640 lbs.

Table N shows the several readings.

Beam XXXIX was tested with the annular rings as in Fig. 84.

The load was gradually increased until it amounted to 51,400 lbs., when the beam failed by the tearing apart of the fibres on the tension side.

The maximum skin stress corresponding to this load is 2696 lbs. per square inch.

The co-efficient of elasticity, as determined from an increment in the deflection of .175-in. between the loads of 10,000 and 25,000 lbs., is 433,250 lbs.

Table N shows the several readings.

Beams XL and XLI are the two ends of Beam XXXVII which was tested on Feb. 24th, 1893, the central portion of the beam containing the fracture having been cut out.

Beam XL was tested on March 17th with the annular rings as in Fig. 85. The load was gradually increased until it amounted to 53,650 lbs., when the beam failed by the tearing apart of the fibres on the tension side.

The maximum skin stress corresponding to this load is 3311 lbs. per square inch.

The co-efficient of elasticity, as determined by an increment in the deflection of .19-in. between the loads of 12,000 and 26,000 lbs., is 693,090 lbs.

Table N shows the several readings.

The weight of the beam per cubic foot on the day of the test was 36.13 lbs.

Beam XLI was tested on March 17th, 1893, with the annular rings as in Fig. 86. The load upon the beam was gradually increased until it amounted to 40,500 lbs., when it failed by the tearing apart of the fibres on the tension side.

The maximum skin stress corresponding to this load is 2500 lbs. per square inch.

The co-efficient of elasticity, as deduced from an increment in the deflection of .19 in. between the loads of 10,000 lbs. and 22,000 lbs., is 519,820 lbs. per square inch.

Table N shows the several readings.

The weight of the beam on the day of test was 36.13 lbs. per cubic foot.

Beams, XLII and XLVI were cut out of one large piece of square pine made on the Pettewawa, a tributary of the Ottawa, in 1888. The piece was driven over 1300 miles, and lay in water for four years until it was taken out in the fall of 1892 and piled for winter sawing.

This timber was purchased from Messrs. Shearer & Brown.

Beam XLII was tested March 8th, 1893, with the annular rings as in Fig. 87.

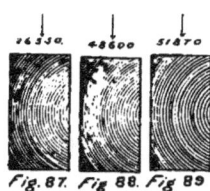

Fig. 87. Fig. 88. Fig. 89.

The load on the beam was gradually increased until it amounted to 26,350 lbs., when the beam failed by the tearing apart of the fibres on the tension side.

The maximum skin stress corresponding to this load is 3815 lbs. per square inch.

The co-efficient of elasticity, as determined by an increment in the deflection of 1.22 ins. between the loads of 2500 lbs. and 13,000 lbs., is 979,220 lbs.

Table O shows the several readings.

The weight of the beam per cubic foot at the date of test was 41.49 lbs.

Beams XLIII and XLIV are the two ends of Beam XLII tested March 8th, the central portion of the beam containing the fracture having been cut out.

Beam XLIII was tested March 31st, with the annular rings as in Fig. 88.

The load was gradually increased until it amounted to 48,600 lbs., when the beam failed by the tearing apart of the fibres on the tension side.

The maximum skin stress corresponding to this load is 3000 lbs. per square inch.

The co-efficient of elasticity, as determined by an increase in the deflection of .19-in. between the loads of 10,000 and 25,000 lbs., is 649,780 lbs. per square inch.

Table O shows the several readings.

Beam XLIV was tested March 31st, 1893, with the annular rings as in Fig. 89.

The load upon the beam was gradually increased until it amounted to 51,870 lbs., when it failed by the tearing apart of the fibres on the tension side.

The maximum skin stress corresponding to this load is 3148 lbs. per square inch.

The co-efficient of elasticity, as determined by an increment in the deflection of .19-in. between the loads of 1000 and 25,000 lbs., is 649,780 lbs. per square inch, the same co-efficient as in beam XLIII.

Table O shows the several readings.

Beam XLV was tested March 11th, 1893, with the annular rings as in Fig. 90.

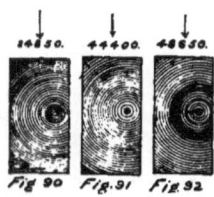

The load upon the beam was gradually increased until it amounted to 24,850 lbs., when it failed by the tearing apart of the fibres on the tension side.

The maximum skin stress corresponding to this load is 3681 lbs. per square inch.

The co-efficient of elasticity, as determined from an increment in the deflection of .81-in. between the loads of 2500 and 12,000 lbs., is 956,540 lbs.

Table P shows the several readings.

Beams XLVI and XLVII are the two ends of Beam XLV, tested on March 11th, 1893, the central portion containing the fracture having been cut out.

Beam XLVI was tested March 30th, 1893, with the annular rings as in Fig. 91.

The load upon the beam was gradually increased until it amounted to 44,400 lbs., when it failed by the tearing apart of the fibres on the tension side.

The maximum skin stress corresponding to this load is 2740 lbs. per square inch.

The co-efficient of elasticity, as determined by an increment in the deflection of .23-in. between the loads of 10,000 and 25,000 lbs., is 536,770 lbs.

Table P shows the several readings.

Beam XLVII was tested March 30th, 1893, with the annular rings as in Fig. 92.

The load upon the beam was gradually increased until it amounted to 48,650 lbs., when it failed by the tearing apart of the fibres on the tension side.

The maximum skin stress corresponding to this load is 3003 lbs. per square inch.

The co-efficient of elasticity, as determined by an increment in the deflection of .2-in. between the loads 10,000 and 25,000 lbs., is 617,283 lbs.

Table P shows the several readings.

Beams XLVIII to L were sent to the laboratory by Mr. P. A. Peterson. These beams were purchased from the Pembroke Lumber Company, and are supposed to have been similar in quality to the timber used on the Pembroke section of the Canadian Pacific Railway.

Beam XLVIII was tested March 1st, 1894, with the annular rings as in Fig. 93. The darkened portion, Fig. 96, represents sapwood.

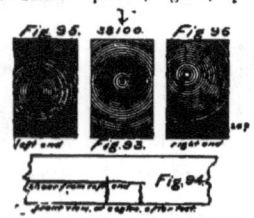

The load upon the beam was gradually increased until it amounted to 38,100 lbs., when the beam failed by the crippling of the material at the support on the compression side, Fig. 94. The load was still gradually increased until it amounted to 47,960 lbs., when a complete fracture took place by the tearing apart of the fibres on the tension side at the centre, and simultaneously by a longitudinal shearing throughout one-half of the length of the beam, as in Figs. 94, 95.

The maximum skin stress corresponding to the breaking load of 38,100 lbs. is 3991 lbs. per square inch; the maximum skin stress corresponding to the load of 47,960 lbs. is 5017 lbs. per square inch.

The total compression of the timber at the centre was .93-in., so that, taking the effective depth to be 14.3875 ins., the maximum compressive skin stress at the support would be 4161 lbs. per square inch, the corresponding maximum tensile skin stress being 4652 lbs. per square inch.

Assuming the usual law to hold good for the whole of the effective depth, the maximum skin stress would be 4447 lbs. per square inch.

The co-efficient of elasticity, as determined by an increment in the deflection of .375-in., between the loads of 2000 lbs. and 19,000 lbs., is 1,164,700 lbs.

Table Q gives the several readings.

The total weight of the beam on March 1st, the date of test, was 524 lbs. 10 ozs., or 41.08 lbs. per cubic foot, and on February 1st the weight was 597 lbs., or 46.73 lbs. per cubic foot, showing a loss of weight at the rate of .209-lb. per cubic foot per day.

The time occupied by the test was 48 minutes.

Beam XLIX was tested March 2nd, 1894, with the annular rings as in Fig. 97. The darkened portions represent sapwood.

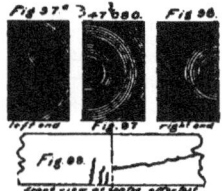

The load upon the beam was gradually increased until it amounted to 47,080 lbs., when the beam failed by the tearing apart of the fibres on the tension side, accompanied simultaneously by a longitudinal shear and a crippling of the material in the compression side, Figs. 98, 99.

The maximum skin stress corresponding to the breaking load is 4936 lbs. per square inch.

The total compression of the material at the centre was 2.8 ins., so that taking 13.095 ins. as the effective depth, the maximum compressive stress would be 5156 lbs. per square inch, and the corresponding skin tensile stress would be 7353 lbs. per square inch.

Assuming the usual law to hold good for the whole of the effective depth, 6835 lbs. per square inch would be the maximum skin stress.

The co-efficient of elasticity, as determined by an increment of .435-in., between the loads of 3000 and 21,000 lbs., is 1,052,600 lbs.

Table Q shows the several readings.

The weight of the beam was 525 lbs. 12 ozs., or 41.33 lbs. per cubic foot February 1st, and 473 lbs. 12 ozs., or 37.24 lbs. per cubic foot on March 2nd, showing a loss of weight at the rate of .141-lbs. per cubic foot per day.

The time occupied by the test was fifty minutes.

Beam L was tested March 10th, 1894, with the annular rings as in Fig. 100.

The load upon the beam was gradually increased until it amounted to 32,200 lbs., when it failed by the tearing apart of the fibres on the tension side.

The maximum skin stress corresponding to this load is 4370 lbs. per square inch.

The co-efficient of elasticity, as deduced from an increment in the deflection of .805-in., between the loads of 1000 and 19,000 lbs., is 1,184,240 lbs.

Table Q shows the several readings.

The weight of the beam was 509 lbs. 12 ozs. or 33.64 lbs. per cubic foot, on March 10th, the date of test, and 575 lbs. 8 ozs., or 37.25 lbs. per cubic foot, on February 1st, showing a loss of weight at the rate of .0975-lb. per cubic foot per day.

OLD WHITE PINE.

Beams LI to LIII are three old white pine stringers sent to the laboratory by Mr. P. A. Peterson. These stringers had been in service since 1885, i.e., for about eight years; they were removed from the trestles during the summer of 1892.

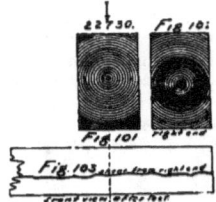

Beam LI was tested December 1st, 1893, with the annular rings as in Fig. 101.

The load upon the beam was gradually increased until it amounted to 22,730 lbs. when the beam failed by shearing, longitudinally as in Figs. 102, 103, the distance between the portions of the beam above and below the plane of shear being ¼ in.

The maximum skin stress corresponding to this load is 3212 lbs. per square inch.

The co-efficient of elasticity, as determined by an increment in the deflection of .55-in., between the loads of 2500 lbs. and 12,000 lbs. is 982,480 lbs.

Table R shows the several readings.

The total weight of the beam on December 1st, date of test, was 445 lbs., or 28.3 lbs. per cubic foot. The weight of a length of 14 ft. 1¾ ins. was 376 lbs., or 28.12 lbs. per cubic foot on December 2nd, and 367 lbs. 5 ozs., or 27.47 lbs. per cubic foot on December 8th, showing a loss of weight at the rate of .1083-lb. per cubic foot per day.

Beam LII was tested December 9th, 1893, with the annular rings as in Fig. 104.

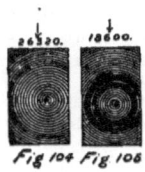

The load upon the beam was gradually increased until it amounted to 26,320 lbs., when the beam failed by the tearing apart of the fibres on the tension side.

The maximum skin stress corresponding to this breaking load is 3589 lbs. per square inch.

The total compression of the material at the support was .37-in., so that, taking 14.85 ins. as the effective depth, the maximum skin compressive stress is 3671 lb. per square inch, the corresponding maximum tensile stress being 3863-lb. per square inch. Assuming the usual law to hold good for the whole of the depth, the maximum skin stress per square inch would be 3774 lbs.

The co-efficient of elasticity, as determined from an increment in the deflection of .635-in. between the loads of 2500 lbs. and 14,500 lbs., is 929,690 lbs.

Table R shows the several readings.

The weight of the beam on November 29th was 430 lbs., or 28.71 lbs. per cubic foot, and on December 9th, the date of test, the weight was 415 lbs. 6½ ozs., or 26.08 lbs. per cubic foot, showing a loss of weight at the rate of .263-lb. per cubic foot per day.

Beam LIII was tested December 9th, 1893, with the annular rings as in Fig. 105.

The beam was a poor specimen, being full of knots and season cracks, and partly decayed. The grain on the top was parallel, while on the sides it was somewhat oblique.

The load upon the beam was gradually increased until it amounted to 18,600 lbs., when it failed by the tearing apart of the fibres on the tension side.

The maximum skin stress due to this breaking load is 2495 lbs. per square inch.

The co-efficient of elasticity, as determined by an increment in the deflection of .55-in. between the loads of 1500 lbs. and 10,000 lbs., is 650,930 lbs.

Table R shows the several readings.

The weight of the beam was 450 lbs. 12 ozs., or 29.02 lbs. per cubic foot on Nov. 9th, and 438 lbs. 13 ozs., or 28. 25 lbs. per cubic foot on Dec. 8th, showing a loss of weight at the rate of .0855 lb. per cubic foot per day.

The time occupied by the test was 20 minutes.

The following Table gives the summary of the results obtained for White Pine:—

Beams.	Dimensions in inches.			Weight in lbs. per cubic foot at date of test.	Maximum skin stress in lbs. per sq. in.	Co-efficient of Elasticity in lbs.
	l	d	b			
NEW TIMBER.						
XLII.	288 × 18		× 9	41.49	3,815	979,220
XLV.	288 × 18		× 9	41.49	3,681	956,540
XLVIII.	150 × 15.1875		× 9.375	41.08	3,991	1,164,700
XLVI.	120 × 18		× 9	39.53	2,740	536,770
XLVII.	120 × 18		× 9	39.40	3,003	617,283
XLIII.	120 × 18		× 9	39.50	3,000	619,780
XLIV.	120 × 18		× 9	39.40	3,148	619,780
XXXVI.	288 × 18		× 9	37.25	2,993	500,000
XLIX.	150 × 15.375		× 9.125	37.24	1,936	1,052,600
XXXVII.	288 × 18		× 9	36.43	3,555	
XL.	120 × 18		× 9	36.13	3,311	693,090
XLI.	120 × 18		× 9	36.13	2,500	519,820
XXXVIII.	114 × 18		× 9	34.78	3,075	622,640
XXXIX.	102 × 18		× 9	34.78	2,696	433,250
L.	186 × 15		× 9.0625	33.64	4,370	1,184,240
OLD TIMBER.						
LIII.	180 × 15		× 9.05	28.25	2,495	650,930
LI.	192 × 15.12		× 9	28.3	3,212	982,480
LII.	180 × 14.85		× 9.05	26.08	3,589	929,690

Hence, for the new timber,

The average weight in lbs. per cubic foot = 37.88.
    " co-efficient of elasticity in lbs. per sq. in. = 754,265.
    " maximum skin stress " " = 3388.

The following data are suggested for practice:—

The average weight in lbs. per cubic foot = 37.8.
" co-efficient of elasticity in lbs. per sq. in. = 754,000.
" maximum skin stress " " = 3,300.
" safe working skin stress in lbs. per sq. in., 3 being at factor of safety = 1100.

Further experiments will probably show that these data require some modification. In fact, the actual skin stress and co-efficients of elasticity are certainly greater than those given in the preceding table, which have been calculated on the assumption that the amount of the compression at the central support is sufficiently small to be disregarded, but it has been shewn, as for example, in the case of Beam XLIX, that the skin stresses are largely affected by this compression. The co-efficients of elasticity are also necessarily increased by the diminution in the effective depth. Similar remarks apply to the other timbers.

From the experiments with the old White Pine stringers, it might be inferred that these timbers have lost considerably in weight, but that they have in a great degree retained their strength and stiffness. Other old Timbers will require to be tested, however, before any definite statement can be made on the subject.

NEW SPRUCE BEAMS.

Beam LIV was tested Nov. 2nd, 1893, with the annular rings as in Fig. 106.

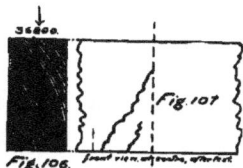

This stick was sent to the laboratory by Mr. T. J. Claxton. It was cut out of a tree felled near the Skeena River, British Columbia, on the Pacific Coast, about six hundred miles north of Victoria. The log was felled in Dec., 1892, or January, 1893, and was over 100 ft. in length, squared 36 ins. at the small end, and would have provided from 12,000 to 15,000 of market lumber.

The beam in question was sawn from the log in June, 1893, and was shipped by steamer at the end of June from the town of Claxton, situated at the mouth of the Skeena River, where the mills are located. At Victoria the beam was transhipped and brought down in August via the C.P.R. to Montreal. It was delivered at the laboratory early in September.

It might, perhaps, be of interest to note that the cost of freight for this beam from Claxton to Victoria was $4.00; from Victoria to Vancouver $2.00; from Vancouver to Montreal $46.00; and the cartage to the University $4 00, making a total cost of freight of $56.00.

It is said that the spruce from the Skeena District is of a specially fine quality, having a clear straight grain, and possessing a large amount of toughness.

The load upon the beam was gradually increased until it amounted to 36,800 lbs., when the beam failed by the crippling of the fibres on the compression side, Fig. 107.

The maximum skin stress corresponding to this breaking load 5908 lbs. per square inch.

The total compression of the material at the central support was .5 in., so that taking the effective depth as 17 ins., the maximum skin compressive stress is 5941 lbs. per square inch, the corresponding skin tensile stress being 6301 lbs. per square inch.

If it is assumed that the usual law holds good for the whole of the effective depth of 17 ins., the maximum skin stress is 6260 lbs. per square inch.

The co-efficient of elasticity, as deduced from an increment in the deflection of 1.15 ins. between the loads of 1000 and 15,000 lbs., is 1,528,499 lbs.

Table S shows the several readings.

The weight of the beam on Oct. 3rd was 751 lbs. 6 ozs., or 27.206 lbs per cubic foot, and on Nov. 3rd, the date of test, it weighed 735 lbs. 2½ ozs., or 26.614 lbs. per cubic foot, showing a loss while in the laboratory at the rate of .019 lbs. per cubic foot per day.

Beams LV and LVI are the ends of Beam LIV, the central portion containing the fracture having been cut out.

Beam LV was tested Nov. 3rd, 1893, with the annular rings as in Fig. 108.

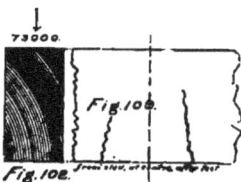

The load was gradually increased until it amounted to 73,000 lbs., when it failed by the crippling of the fibres on the compression side Fig. 109.

The maximum skin stress corresponding to this load is 4839 lbs. per square inch.

The maximum compression of the material at the central support was 2 ins., so that taking 15 5 ins. as the effective depth, the maximum compressive skin stress is 5123 lbs. per square inch, the corresponding tensile skin stress being 6641 lbs. per square inch.

If it is assumed that the usual law holds good for the whole of the effective depth, the maximum skin stress becomes 6176 lbs.

As soon as the beam was relieved of load, the amount of compression at the support was immediately diminished by .9-in., and at the end of thirteen days the amount of compression was .82 in.

The co-efficient of elasticity, as determined by an increment in the deflection of .17-in., between the loads of 3000 lbs. and 10,000 lbs., is 1,070,950 lbs.

Table T shows the several readings.

The weight of the beam on Nov. 3rd, date of test, was 26.614 lbs. per cubic foot

Beam LVI was tested Nov. 4th, 1893, with the annular rings as in Fig. 110.

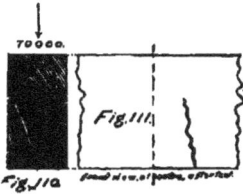

The load was gradually increased until it amounted to 70,000 lbs. when it failed by the crippling of the fibres on the compression side Fig. 111.

The maximum skin stress corresponding to this breaking load is 4614-lbs. per square inch.

The maximum compression at the centre of support was 1.9 ins., so that taking 15.6 ins. as the effective depth, the maximum compressive skin stress is 4916 lbs. per square inch, the corresponding tensile skin stress being 6280 lbs. per square inch.

If it is assumed that the usual law holds good for the whole of the effective depth, then the maximum skin stress becomes 5806 lbs. per square inch.

Ten days after this beam had been relieved of load, the amount of the compression of the timber at the centre of support was diminished to .77 in.

The co-efficient of elasticity, as determined by an increment in the deflection of .18 in. between the loads of 10,000 lbs. and 30,000 lbs., is 1,011,450 lbs.

Table T shows the several readings.

The weight of this beam on Nov. 3rd was 26.014 lbs. per cubic foot.

OLD SPRUCE.

Beams LVII-LIX were three spruce stringers sent to the laboratory by Mr. P. A. Peterson.

Beams LVII and LVIII were cut at Galbraith's Mill, three miles from Sherbrooke, in 1886, and grew near the same place. They were used in the construction of the bridge near Lennoxville in the winter of 1886-87, and had been in service until the summer of 1894, or for a period of about eight years.

Beam LIX was taken out of Bridge F 61 at Roxton Falls during the summer of 1894, and had been in service since 1885, i.e., for about eight years. This stringer was purchased by Bridge-master MacFarlane, and no further information has been obtained as to its history. The stringer was boxed ½-in. at the ends on the bearings, and several season cracks were shown on the surface.

Beam LVII was tested on the 21st April with the annular rings as in Fig. 112.

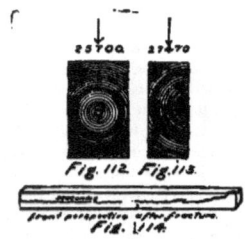

Fig. 112. Fig. 113.

Fig. 114.

The load upon the beam was gradually increased until it amounted to 25,700 lbs., when the beam failed by shearing longitudinally along the surface of a season crack, the distance between the portions above and below the plane of shear at the end being ⅜-in.

Immediately after the fracture the jockey weight was run back until the lever again floated, the load upon the beam being 21,000 lbs. This load was then gradually increased until it amounted to 24,700 lbs., when failure occurred by the tearing apart of the fibres on the tension side and by a further crippling of the fibres on the compression side. The lap at the end of the plane of shear was also increased to ⅝-in.

The maximum skin stress corresponding to the breaking load of 25,700 lbs. is 3459 lbs. per square inch.

The maximum compression of the material at the support was .31-in., so that taking the effective depth to be 14.69 ins., the maximum compressive skin stress is 3526 lbs. per square inch, the corresponding tensile skin stress being 3678 lbs. per square inch.

If it is assumed that the usual law holds good for the whole of the effective depth, then the maximum skin stress becomes 3007 lbs. per square inch.

The co-efficient of elasticity, as determined by an increment in the deflection of .7-in. between the loads of 1500 and 12,500 lbs., is 1,123,400 lbs.

Table U shows the several readings.

The weight of this beam on April 10th was 502 lbs., or 33.82 lbs. per cubic foot; its weight on April 21st, date of test, was 491 lbs. 4 ozs., or 33.09 lbs. per cubic foot, showing a loss of weight at the rate of .0645 lbs. per cubic foot per day.

Beam LVIII was tested May 1st, 1894, with the annular rings as in Fig. 113. Season cracks ran intermittently from end to end of the beam.

The load upon this beam was gradually increased until it amounted to 27,470 lbs. Under this load the beam failed by shearing longitudinally along a season crack, as shown in Fig. 114, with a partial tension fracture near the end of the beam. The season crack for a distance of about 3 ft. from the centre of the beam appears weathered through the entire thickness of the beam.

Previously, however, to this longitudinal shear, the beam had evidently failed by the crippling of the material, Fig. 114, on the compression side along a line near the centre of the beam where the timber was apparently free from knots and where the fibres were parallel with the axis.

The maximum skin stress corresponding to the load of 27,470 lbs. is 5709 lbs. per square inch.

The co-efficient of elasticity, as determined by an increment in the deflection of .575-lbs. between the loads of 2000 and 12,000 lbs., is 1,316,900 lbs.

Table U shows the several readings.

The weight of the beam on March 10th was 267 lbs. 1 oz., or 27.36 lbs. per cubic foot, and its weight on May 2nd was 258 lbs. 6 ozs., or 26.47 lbs. per cubic foot, showing a loss of weight while in the laboratory at the rate of .0168 lb. per cubic foot per day.

Beam LIX was tested June 2nd, 1894, with the annular rings as in Fig. 115.

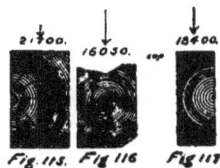

Fig. 115. Fig. 116. Fig. 117.

The load was gradually increased until it amounted to 21,700 lbs., when the beam failed by the tearing apart of the fibres on the tension side.

The maximum skin stress corresponding to this load is 2963 lbs. per square inch.

The maximum compression at the centre was .7-in., so that taking 14.3 ins. as the effective depth, the maximum compressive skin stress is 3079 lbs. per square inch, the corresponding tensile skin stress being 3396 lbs. per square inch.

If it is assumed that the usual law holds good for the whole of the effective depth, then the maximum skin stress is 3261 lbs. per sq. in.

The co-efficient of elasticity, as determined by an increment in the deflection of .43-in. between the loads of 2000 lbs. and 10,000 lbs., is 905,601 lbs.

Table U shows the several readings.

The weight of the beam on June 1st was 415 lbs. 13 ozs., or 30.12 lbs. per cubic foot. Its weight on June 8th was 440 lbs , or 29.72 lbs. per cubic foot, showing a loss of weight at the rate of .0571-lb. per cubic foot per day.

Beams LX and XLI are two old spruce stringers sent to the laboratory by Mr. P. A. Peterson.

They had been in use in Culvert E 39 on the north division of the South Eastern Railway, 1½ miles north of Waterloo Station, since Oct., 1891, or for about three years.

These timbers were cut and sawn at Keene & Company's mills at the boundary east of Megantic.

Beam LX was tested on Nov. 10th, 1894, with the annular rings as in Fig. 116.

The upper portion of the stringer, i.e., the part in tension, was partially rotten to a depth of about 1-in., and the effective depth at the centre of the beam did not exceed 11¼ ins. The remainder of the section at the centre was in a perfectly sound and good condition.

The load upon the beam was gradually increased until it amounted to 16,050 lbs., when it failed by the tearing apart of the fibres on the tensile side. The load was still increased, and a more complete fracture occurred under a load of 21,240 lbs. Immediately after this second fracture the jockey weight was run back until the lever again floated, when the load was 15,900 lbs. The load was again gradually increased until it amounted to 18,800 lbs., when fracture again occurred.

The maximum skin stress corresponding to the breaking load of 16,050 lbs. is 2934 lbs.

The maximum compression of the material at the centre was .25-in., so that taking the effective depth to be 11. ins., the maximum compressive skin stress is 3043 lbs. per square inch, and the corresponding tensile skin stress is 3184 lbs. per square inch.

If it is assumed that the usual law holds good for the whole of the effective depth, the maximum skin stress becomes 3118 lbs. per square inch.

The co-efficient of elasticity, as determined by an increment in the deflection of .390-in. between the loads of 2000 and 12,000 lbs., is 1,352,250 lbs. per square inch.

Table V gives the several readings.

The weight of this beam on Nov. 10th, date of test, was 255 lbs. 12½ ozs., or 27.26 lbs. per cubic foot.

Beam LXI was tested Nov. 17th, 1894, with the annular rings as in Fig. 117. There were season cracks from end to end on the front face and numerous knots of medium and small size on the sides. The darkened portion indicates sapwood.

The load upon the beam was gradually increased until it amounted to 18,400 lbs., when the beam failed by the tearing apart of the fibres on the tension face.

The maximum skin stress corresponding to this load is 4309 lbs. per square inch.

The maximum compression of the material at the centre was .21 in., so that taking the effective depth to be 14.29 ins., the maximum skin compressive stress is 4432 lbs. per square inch, the corresponding tensile skin stress being 4565 lbs. per square inch.

If it is assumed that the usual law holds good for the whole of the effective depth, the maximum skin stress becomes 4502 lbs. per square inch.

The co-efficient of elasticity, as determined from an increment of .6-in. in the deflection between the loads of 1000 lbs. and 9000 lbs., is 1,250,850 lbs.

The weight of this beam on Nov. 17th, date of test, was 267 lbs., or 28.85 lbs. per cubic foot.

The following Table gives a summary of the results obtained for Spruce :—

NEW TIMBER.

Beam.	Dimensions in inches.			Weight in lbs. per cubic foot at date of test.	Maximum skin stress in lbs. per sq. in.	Co-efficient of Elasticity in lbs.
	l	d	b			
LIV.	288 ×	17.5 ×	8.875	26.614	5,908	1,528,499
LV.	120 ×	17.5 ×	8.875	26.614	4,839	1,070,950
LVI.	120 ×	17.5 ×	9.9475	26.614	4,614	1,011,450

OLD TIMBER.

LVII.	180 ×	15 ×	9	33.09	3,459	1,123,400
LIX.	180 ×	15 ×	9	30.12	2,963	905,601
LXI.	186 ×	14.5 ×	5.625	28.85	4,309	1,250,850
LX.	138 ×	11.25 ×	8.875	27.26	2,934	1,352,250
LVIII.	180 ×	14.75 ×	6	26.47	5,709	1,316,900

Beams LV and LVI were cut out of Beam LIV as already described. The wide variation in the value of the skin-stress and of the co-efficient of elasticity is undoubtedly due to the fact that the amount of the compression at the central support has been disregarded in the calculations. If this compression is taken into account, and if it is assumed that the ordinary theory of flexure holds good for the whole of the effective depth, it has been shewn that the skin-stresses in lbs. per sq. in. become 6260 for Beam LIV, 6176 for Beam LV, and 5806 for Beam LVI, the variation in the magnitude of the stresses being comparatively small.

Further experiments will be made with new spruce beams.

The old spruce stringers were found to possess ample strength and stiffness for the work they were designed to do. The experiments gave :—

29.15 lbs. as the average weight per cubic foot.
1,189,800 "   "   co-efficient of elasticity.
3875 "   "   maximum skin-stress per sq. in.

The following Tables A to V give the end deflections and in some cases the deflections at points dividing the beam into four, six, or eight equal parts, the distances of these points from the ends being stated at the heads of the columns.

Tables A to I show the deflections in inches of Canadian New Douglas Fir Beams (I to XXV) under gradually increased loads.

TABLE A.

Deflections of Beam I at ends.

Loads in lbs.	Deflection.	Loads in lbs.	Deflection.	Loads in lbs.	Deflection.	Loads in lbs.	Deflection.	Loads in lbs.	Deflection.
2,000	.02	9,000	.095	16,000	.18	23,000	.27	30,000	.39
2,500	.03	9,500	.10	16,500	.19	23,500	.28	30,500	.40
3,000	.03	10,000	.11	17,000	.195	24,000	.285	31,000	.41
3,500	.035	10,500	.115	17,500	.20	24,500	.295	31,500	.42
4,000	.04	11,000	.12	18,000	.205	25,000	.30	32,000	.43
4,500	.045	11,500	.125	18,500	.21	25,500	.31	32,500	.445
5,000	.05	12,000	.13	19,000	.22	26,000	.315	33,000	.46
5,500	.055	12,500	.14	19,500	.225	26,500	.32	34,000	.49
6,000	.06	13,000	.145	20,000	.230	27,000	.33	35,000	.51
6,500	.07	13,500	.15	20,500	.24	27,500	.34	36,000	.53
7,000	.075	14,000	.155	21,000	.245	28,000	.35	37,000	.56
7,500	.075	14,500	.16	21,500	.25	28,500	.36		
8,000	.08	15,000	.165	22,000	.255	29,000	.37		
8,500	.09	15,500	.17	22,500	.265	29,500	.38		

Breaking weight of Beam I = 45,000 lbs.

TABLE B.

Loads in lbs.	Deflections of Beams.						
	II	III	IV	V	VI	VII	VIII
	Ends.	Ends.	Ends.	Ends.	Ends.	Ends.	Ends.
300							.02
500		.05		.005	.02	.015	.03
800							.05
1,000		.08	.03	.01	.04	.03	.07
1,300							.09
1,500		.11	.045	.02	.06	.04	.10
1,800							.12
2,000	.035	.14	.05	.03	.075	.06	.135
2,200							.15
2,400							.165
2,500		.155	.055	.05	.10	.075	
2,600							.18
2,800							.195
3,000		.18	.065	.055	.12	.10	.205
3,400							.235
3,500		.21	.08	.065	.14	.115	
3,800							.26
4,000	.05	.23	.095	.07	.16	.125	.28
4,500		.25	.105	.08	.18	.14	.315
5,000			.115	.09	.20	.155	.35
5,500			.13	.105	.22	.175	.39
6,000	.065		.145	.11	.24	.195	
6,500			.155	.125	.26	.21	
7,000			.165	.135	.28	.22	
7,500			.18	.145	.305	.235	
8,000	.075		.19	.16	.32	.25	
8,500			.20	.17		.27	
9,000			.215	.18			
9,500			.23	.195			
10,000	.085		.245	.205			
10,500			.26	.22			
11,000			.28	.235			
11,500			.30	.25			
12,000	.10		.315	.26			
12,500			.33	.27			
13,000	.105		.35	.28			
13,500			.365	.29			
14,000	.110		.38	.305			
14,500				.315			
15,000	.115			.33			
15,500				.345			
16,000	.12						
16,400						.75	
17,000	.13						
18,000	.135						
20,000	.14						
21,000				.72			
22,000	.15						
24,000	.165						
26,000	.175						
28,000	.190						

Breaking Weight of Beam II = 36,575 lbs.
" " III = 12,950 "
" " IV = 16,720 "
" " V = 23,610 "
" " VI = 15,480 "
" " VII = 17,615 "
" " VIII = 11,700 "

TABLE C.

Loads in lbs.	Deflections of Beam IX.					Deflections of Beam X.				
	34 ins.	68 ins.	Ends.	68 ins.	34 ins.	33 ins.	66 ins.	Ends.	66 ins.	33 ins.
1000	.01	.01	.02	.01	.01	.02	.01	.02	.01	.02
1500	.03	.02	.04	.02	.03	.05	.02	.05	.02	.05
2000	.03	.03	.05	.025	.04	.07	.03	.08	.04	.07
2500	.04	.03	.05	.03	.05	.10	.05	.11	.05	.10
3000	.10	.07	.06	.05	.09	.12	.06	.14	.06	.12
3500	.10	.08	.12	.05	.10	.15	.07	.17	.07	.15
4000	.10	.08	.13	.055	.10	.17	.09	.20	.08	.17
4500	.10	.08	.14	.065	.11	.20	.10	.23	.10	.20
5000	.15	.10	.18	.085	.15	.22	.11	.26	.115	.22
5500	.15	.11	.19	.09	.16	.25	.12	.29	.12	.25
6000	.15	.12	.20	.10	.17	.27	.14	.32	.14	.27
6500	.19	.13	.24	.11	.20	.30	.15	.35	.15	.30
7000	.20	.13	.25	.115	.20	.32	.17	.38	.16	.32
7500	.20	.13	.25	.11	.21	.35	.18	.41	.18	.35
8000	.20	.13	.26	.125	.22	.37	.20	.44	.20	.37
8500	.22	.14	.27	.135	.24	.40	.21	.47	.21	.40
9000	.22	.15	.28	.11	.24	.42	.22	.50	.22	.42
9500	.22	.15	.28	.145	.25	.45	.23	.53	.23	.45
10000	.26	.16	.33	.16	.28	.47	.25	.56	.24	.47
10500	.33	.20	.40	.19	.34	.49	.26	.58	.25	.49
11000	.34	.21	.42	.20	.35	.51	.27	.61	.27	.51
11500	.35	.22	.44	.205	.36	.54	.29	.64	.29	.54
12000	.39	.23	.47	.22	.40	.56	.30	.68	.30	.56
12500	.40	.24	.49	.22	.40	.59	.32	.71	.32	.59
13000	.40	.24	.50	.23	.41	.61	.33	.74	.33	.61
13500	.45	.27	.54	.25	.45	.64	.34	.77	.34	.64
14000	.45	.27	.55	.255	.46	.66	.36	.80	.36	.66
14500	.45	.27	.56	.26	.46	.69	.37	.83	.375	.69
15000	.50	.29	.60	.27	.50	.71	.39	.86	.39	.71
15500	.50	.30	.61	.28	.51	.74	.40	.89	.40	.74
16000	.50	.30	.62	.29	.52	.75	.41	.92	.41	.76
16500	.55	.31	.66	.31	.55	.79	.43	.96	.43	.79
17000	.55	.32	.67	.31	.56	.81	.44	.99	.45	.82
17500	.56	.33	.68	.32	.57	.85	.46	1.02	.46	.85
18000	.56	.33	.69	.325	.58					
18500	.60	.36	.75	.35	.62					
19000	.63	.36	.77	.35	.64					
19500	.64	.37	.78	.36	.65					
20000	.65	.37	.79	.365	.66					
40000			1.75							
47000			2.20							

Breaking Weight of Beam IX = 51,600 lbs.
" " " X = 18,000 "

TABLE D.

Loads in lbs.	Deflections of Beam XI.					Deflections of Beam XII.				
	34 ins.	68 ins.	Ends.	68 ins.	34 ins.	34 ins.	68 ins.	Ends.	68 ins.	34 ins.
1000	......	......	......	......	......	.01	.005	.01	.01	.01
1500	.02	.01	.035	.015	.025	.03	.02	.035	.02	.035
2000	.05	.02	.05	.025	.04	.05	.025	.055	.03	.05
2500	.06	.03	.075	.035	.06	.065	.04	.075	.05	.07
3000	.075	.04	.10	.045	.08	.09	.045	.10	.05	.09
3500	.10	.05	.115	.055	.095	.105	.06	.12	.06	.105
4000	.11	.06	.135	.06	.11	.12	.07	.145	.07	.12
4500	.13	.07	.16	.07	.135	.15	.075	.165	.08	.145
5000	.15	.075	.175	.075	.14	.155	.09	.185	.09	.155
5500	.16	.085	.20	.09	.16	.17	.10	.205	.10	.17
6000	.185	.10	.22	.10	.18	.19	.11	.23	.11	.19
6500	.20	.105	.24	.11	.195	.21	.12	.25	.12	.21
7000	.215	.115	.26	.11	.215	.23	.13	.27	.13	.235
7500	.24	.125	.28	.13	.235	.25	.14	.295	.14	.25
8000	.25	.135	.30	.14	.245	.27	.15	.315	.15	.27
8500	.26	.145	.32	.15	.265	.29	.15	.34	.16	.29
9000	.27	.15	.33	.155	.27	.305	.17	.36	.17	.305
9500	.30	.16	.35	.165	.29	.32	.18	.305	.18	.32
10000	.315	.17	.38	.175	.305	.35	.19	.405	.19	.35
10500	.34	.185	.40	.185	.335	.36	.20	.425	.20	.36
11000	.36	.195	.435	.20	.36	.375	.21	.45	.21	.38
11500	.36	.20	.435	.20	.36	.39	.22	.47	.22	.40
12000	.395	.215	.475	.22	.395	.41	.23	.495	.23	.41
12500	.40	.22	.50	.23	.405	.44	.24	.51	.24	.41
13000	.42	.23	.505	.24	.42	.45	.25	.535	.25	.45
13500	.45	.25	.54	.255	.445	.47	.26	.555	.26	.47
14000	.46	.255	.56	.265	.46	.49	.27	.58	.27	.49
14500	.48	.265	.57	.275	.475	.50	.28	.60	.28	.505
15000	.50	.275	.60	.28	.50	.52	.29	.62	.30	.52
15500	.515	.285	.62	.29	.515	.55	.30	.645	.305	.55
16000	.535	.295	.645	.30	.53	.555	.305	.665	.31	.56
16500	.54	.30	.65	.30	.535	.575	.32	.69	.32	.57
17000	.58	.32	.695	.32	.575	.60	.325	.71	.33	.60
17500	.585	.32	.70	.325	.575	.61	.33	.73	.345	.615
18000	.61	.34	.735	.345	.61	.63	.345	.755	.35	.635
18500	.61	.34	.745	.35	.615	.65	.35	.77	.36	.65
19000	.65	.36	.78	.365	.655	.665	.36	.80	.375	.665
19500	.65	.36	.785	.375	.655	.685	.37	.82	.385	.69
20000	.655	.365	.80	.375	.66	.705	.38	.85	.40	.705
20500	......	......	......	......	......	.73	.395	.87	.41	.725
21000	......	......	......	......	......	.75	.40	.89	.415	.75
21500	......	......	......	......	......	.75	.405	.90	.415	.75
22000	......	......	......	......	......	.78	.42	.935	.435	.78
22500	......	......	......	......	......	.81	.435	.96	.45	.805
23000	......	......	......	......	......	.82	.445	.98	.455	.82
24000	......	......	.94	......	......	......	......	......	......	......
26500	......	......	......	......	......	......	......	1.12	......	......
28000	......	......	1.14	......	......	......	......	1.17	......	......
29000	......	......	......	......	......	......	......	1.22	......	......
32000	......	......	......	......	......	......	......	1.40	......	......
33000	......	......	1.35	......	......	......	......	1.42	......	......
35800	......	......	1.45	......	......	......	......	......	......	......
37000	......	......	......	......	......	......	......	1.67	......	......
39000	......	......	......	......	......	......	......	1.97	......	......
42000	......	......	......	......	......	......	......	2.00	......	......
45000	......	......	......	......	......	......	......	2.28	......	......
48000	......	......	......	......	......	......	......	2.73	......	......
49000	......	......	......	......	......	......	......	2.9	......	......

Breaking Weight of Beam XI = 35,800 lbs.
" " " XII = 49,000 "

TABLE E.

Loads in lbs.	Deflections of Beam XIII.					Deflection of Beam XIV.	Deflections of Beam XV.					Deflection of Beam XVI.
	34 ins.	68 ins.	Ends	68 ins.	34 ins.	Ends	33 ins.	66 ins.	Ends	66 ins.	33 ins.	Ends
760	....	....	....	....	....	.03	....	....	....	....	....	....
1000	.25	.02	.04	.02	.025	.05	.01	.01	.02	.01	.02	.025
1110	....	....	....	....	....	....	....	....	....	....	....	.035
1500	.05	.035	.07	.03	.05	.085	.04	.02	.05	.025	.01	.05
1900	....	....	....	....	....	....	....	....	....	....	....	.075
2000	.08	.05	.105	.05	.08	.115	.055	.035	.08	.045	.06	....
2300	....	....	....	....	....	....	....	....	....	....	....	.09
2500	.10	.065	.14	.065	.11	.15	.08	.045	.095	.05	.075	....
2600	....	....	....	....	....	....	....	....	....	....	....	.10
2800	....	....	....	....	....	.17	....	....	....	....	....	....
3000	.14	.08	.17	.08	.14	.19	.10	.05	.115	.06	.10	.125
3200	....	....	....	....	....	.20	....	....	....	....	....	....
3100	....	....	....	....	....	.22	....	....	....	....	....	.145
3500	.16	.10	.21	.10	.16	....	.11	.065	.11	.07	.12	....
3600	....	....	....	....	....	.225	....	....	....	....	....	....
3800	....	....	....	....	....	.25	....	....	....	....	....	.16
4000	.20	.11	.215	.11	.20	.255	.13	.08	.16	.085	.14	.175
4140	....	....	....	....	....	.275	....	....	....	....	....	.20
4500	.22	.13	.275	.125	.22	....	.155	.095	.185	.095	.16	....
4800	....	....	....	....	....	.315	....	....	....	....	....	.215
5000	.25	.145	.31	.14	.25	.32	.165	.105	.215	.105	.17	.225
5200	....	....	....	....	....	.345	....	....	....	....	....	.23
5400	....	....	....	....	....	.355	....	....	....	....	....	....
5500	.275	.15	.31	.155	.275	....	.19	.11	.21	.115	.20	....
5600	....	....	....	....	....	.36	....	....	....	....	....	.25
5800	....	....	....	....	....	.39	....	....	....	....	....	....
6000	.30	.165	.36	.17	.30	.40	.21	....	.125	.26	.125	.215	.27
6400	....	....	....	....	....	....	....	....	....	....	....	.29
6500	.33	.18	.40	.185	.33	....	.23	.13	.285	.14	.235	....
6600	....	....	....	....	....	.435	....	....	....	....	....	....
6800	....	....	....	....	....	.465	....	....	....	....	....	.31
7000	.36	.20	.44	.20	.36	.485	.255	.145	.31	.15	.255	.325
7200	....	....	....	....	....	.50	....	....	....	....	....	....
7400	....	....	....	....	....	.505	....	....	....	....	....	.34
7500	.38	.215	.47	.22	.39	....	.27	.155	.335	.16	.275	....
7800	....	....	....	....	....	.54	....	....	....	....	....	.36
8000	.41	.225	.50	.23	.41	.56	.295	.165	.35	.175	.30	.375
8300	....	....	....	....	....	.585	....	....	....	....	....	....
8400	....	....	....	....	....	....	....	....	....	....	....	.40
8500	.45	.245	.54	.245	.45	....	.31	.18	.38	.18	.315	....
8600	....	....	....	....	....	.605	....	....	....	....	....	....
8800	....	....	....	....	....	....	....	....	....	....	....	.42
9000	.46	.255	.57	.26	.47	.64	.34	.19	.40	.19	.34	.425
9200	....	....	....	....	....	.66	....	....	....	....	....	....
9400	....	....	....	....	....	....	....	....	....	....	....	.45

TABLE E.—(Continued.)

Loads in lbs.	Deflections of Beam XIII.					Deflections of Beam XIV.	Deflections of Beam XV.					Deflections of Beam XVI.
	34 ins.	68 ins.	Ends	68 ins.	34 ins.	Ends	33 ins.	66 ins.	Ends	66 ins.	33 ins.	Ends
9500	.50	.275	.605	.28	.50	....	.35	.20	.425	.205	.355	....
9600	....	....	....	....	....	.69	....	....	....	....	....	....
9800	....	....	....	....	....	.715	....	....	....	....	....	.475
10000	.52	.29	.64	.295	.53	.73	.37	.21	.44	.21	.375	.485
10200	....	....	....	....	....	.76	....	....	....	....	....	....
10400	....	....	....	....	....	.765	....	....	....	....	....	....
10500	.55	.305	.67	.31	.55	....	.40	.22	.475	.22	.40	....
10600	....	....	....	....	....	.80	....	....	....	....	....	....
10800	....	....	....	....	....	.805	....	....	....	....	....	....
11000	.585	.32	.705	.325	.585	....	.415	.23	.50	.24	.415	.54
11300	....	....	....	....	....	.845	....	....	....	....	....	....
11500	.61	.34	.745	.345	.61	....	.44	.24	.525	.25	.445	.565
11700	....	....	....	....	....	.88	....	....	....	....	....	....
12000	.64	.35	.78	.36	.64	.91	.45	.255	.55	.26	.45	.59
12200	....	....	....	....	....	.935	....	....	....	....	....	....
12400	....	....	....	....	....	.95	....	....	....	....	....	....
12500	.66	.365	.81	.375	.67	....	.47	.265	.57	.27	.465	.61
12600	....	....	....	....	....	.955	....	....	....	....	....	....
12800	....	....	....	....	....	1.00	....	....	....	....	....	....
13000	.70	.385	.845	.395	.70	1.00	.495	.275	.60	.28	.50	.65
13200	....	....	....	....	....	1.02	....	....	....	....	....	....
13500	.725	.40	.885	.41	.735	....	.51	.285	.62	.29	.51	.68
14000	.75	.415	.915	.42	.76	....	.54	.295	.64	.30	.54	.71
14500	.795	.435	.96	.445	.795	....	.55	.305	.66	.31	.55	.73
15000	.81	.45	.99	.46	.82	....	.57	.32	.69	.32	.575	.75
15500	.85	.47	1.025	.475	.85	....	.59	.33	.715	.335	.60	.78
16000	.875	.485	1.065	.49	.875	......	.61	.34	.74	.34	.615	.81
16500	.905	.505	1.10	.515	.915	....	.64	.35	.765	.35	.64	.83
17000	.94	.52	1.135	.525	.94	....	.65	.36	.79	.36	.655	.87
17500	.97	.54	1.18	.545	.975	....	.67	.375	.81	.375	.675	.90
18000	1.00	.55	1.22	.56	1.01	....	.69	.385	.835	.39	.70	.93
18500	1.04	.575	1.265	.58	1.045	....	.71	.395	.86	.40	.71	.95
19000	1.06	.59	1.31	.60	1.07	....	.74	.405	.875	.41	.735	.98
19500	1.1	.615	1.35	.62	1.1	....	.75	.415	.91	.42	.75	1.00
20000	1.14	.63	1.39	.635	1.14	....	.77	.425	.94	.43	.775	1.04
20500	1.165	.65	1.43	.655	1.175	....	....	....	....	....	....	1.07
21000	1.21	.67	1.485	.68	1.22	....	....	....	1.20	....	....	1.10
21500	1.24	.685	1.515	.69	1.25	....	....	....	....	....	....	1.13
22000	1.28	.71	1.57	.715	1.29	....	....	....	....	....	....	1.15
22500	....	....	....	....	....	....	....	....	....	....	....	1.17
23000	....	....	....	....	....	....	....	....	....	....	....	1.20
24000	....	....	1.70	....	....	....	....	....	....	....	....	....
25000	....	....	....	....	....	....	....	....	1.30	....	....	....
26000	....	....	1.88	....	....	....	....	....	....	....	....	....
26300	....	....	2.05	....	....	....	....	....	....	....	....	....
27000	....	....	....	....	....	....	....	....	1.45	....	....	....
29000	....	....	....	....	....	....	....	....	1.55	....	....	....
29300	....	....	2.6	....	....	....	....	....	1.70	....	....	....
30000	....	....	....	....	....	....	....	....	1.90	....	....	....
32000	....	....	....	....	....	....	....	....	2.25	....	....	....
35000	....	....	....	....	....	....	....	....	2.33	....	....	....
37000	....	....	....	....	....	....	....	....	....	....	....	....

Breaking weight of Beam XIII = 29,300 lbs.
" " " XIV = 17,600 "
" " " XV = 37,000 "
" " " XVI = 25,580 to 32,000 lbs.

TABLE F.
Deflections of Beams XVII, XVIII and XIX.

Load in lbs.	XVII		XVIII			XIX 1st Loading			Beam gradually relieved of load			2nd loading
	1st Load-ing Ends.	2nd Load-ing E'ds	Beam grad-ually re-lieved of l'd Ends.	3rd Load-ing E'ds		31½ ins.	End	31½ ins.	31½ ins.	End	31½ ins.	
1000	....	.005	.005	.015	....	.005	.010	.010	.010	.015	.020	....
2000	....	.010	.015	.015	....	.020	.045	.030	....	....	....	....
3000	....	.020	.030	.020	....	.050	.060	.055	....	....	....	....
4000	....	.030	.030	.030	....	.060	.090	.070	.055	.095	.060	....
5000	.07	.040	.040	.045	....	.070	.105	.085	....	....	....	....
6000	....	.050	.050	.050	....	.090	.130	.100	....	....	....	....
7000	....	.060	.060	.060	....	.110	.150	.105	....	....	....	....
7500	....	.065	....	....	....	....	....	....	....	....	....	....
8000	....	.075	.070	.070	....	.115	.170	.125	.120	.190	.120	....
8500	....	....	....	....	....	.120	.185	.135	....	....	....	....
9000	....	.080	.085	.075	....	.130	.200	.140	....	....	....	....
9500	....	....	....	....	....	.140	.215	.145	....	....	....	....
10000	.15	.095	.095	.085	....	.150	.225	.150	....	....	....	....
10500	....	.100	....	....	....	.155	.235	.160	....	....	....	....
11000	....	.100	.105	.095	....	.160	.245	.165	....	....	....	....
11500	....	....	....	....	....	.165	.250	.170	....	....	....	....
12000	....	.110	.110	.100	....	.170	.265	.180	.170	.255	.170	....
12500	.18	....	....	....	....	....	....	....	....	....	....	....
13000	....	....	.130	.110	....	.180	.290	.190	....	....	....	....
14000	....	....	.130	.125	....	.200	.310	.205	....	....	....	....
15000	.22	....	.140	.130	....	.210	.330	.220	....	....	....	....
16000	....	....	.150	.150	....	.225	.345	.230	....	....	....	....
17000	....	....	.165	.165	....	....	....	....	....	....	....	....
17500	.26	....	....	....	....	....	....	....	....	....	....	....
18000	....	....	.175	.170	....	....	....	....	....	....	....	....
19000	....	....	.185	.185	....	....	....	....	....	....	....	....
20000	.30	....	....	....	.200	....	....	....	....	....	....	.420
22000	....	....	....	....	....	....	....	....	....	....	....	.465
22500	.34	....	....	....	....	....	....	....	....	....	....	....
24000	....	....	....	....	....	....	....	....	....	....	....	.510
25000	.385	....	....	....	.220	....	....	....	....	....	....	....
26000	.405	....	....	....	....	....	....	....	....	....	....	.550
27000	.425	....	....	....	....	....	....	....	....	....	....	....
28000	.445	....	....	....	....	....	....	....	....	....	....	.600
29000	.465	....	....	....	....	....	....	....	....	....	....	....
30000	.485	....	....	....	.240	....	....	....	....	....	....	.640
31000	.510	....	....	....	....	....	....	....	....	....	....	....
32000	.535	....	....	....	....	....	....	....	....	....	....	.695
33000	.560	....	....	....	....	....	....	....	....	....	....	....
34000	.585	....	....	....	....	....	....	....	....	....	....	....
35000	.61	....	....	....	.310	....	....	....	....	....	....	.730
36000	.64	....	....	....	....	....	....	....	....	....	....	....
37000	.68	....	....	....	....	....	....	....	....	....	....	.780
38000	.715	....	....	....	....	....	....	....	....	....	....	....
39000	.75	....	....	....	....	....	....	....	....	....	....	.830
40000	.795	....	....	....	....	....	....	....	....	....	....	....
41000	.850	....	....	....	.310	....	....	....	....	....	....	.870
42000	.940	....	....	....	....	....	....	....	....	....	....	....
42500	1.005	....	....	....	....	....	....	....	....	....	....	.930
43000	1.030	....	....	....	....	....	....	....	....	....	....	....
43500	1.055	....	....	....	....	....	....	....	....	....	....	....
44000	1.085	....	....	....	.356	....	....	....	....	....	....	....
44500	1.125	....	....	....	....	....	....	....	....	....	....	.980
45000	1.150	....	....	....	....	....	....	....	....	....	....	....
45500	1.240	....	....	....	....	....	....	....	....	....	....	....
46000	1.285	....	....	....	....	....	....	....	....	....	....	1.030
46500	1.315	....	....	....	....	....	....	....	....	....	....	....
47000	1.365	....	....	....	.400	....	....	....	....	....	....	....
47500	1.455	....	....	....	....	....	....	....	....	....	....	....
48000	1.600	....	....	....	.440	....	....	....	....	....	....	1.100
48100	1.640	....	....	....	....	....	....	....	....	....	....	....
48200	1.675	....	....	....	....	....	....	....	....	....	....	....
48300	1.720	....	....	....	....	....	....	....	....	....	....	....
48400	1.830	....	....	....	....	....	....	....	....	....	....	....
48500	1.910	....	....	....	....	....	....	....	....	....	....	....
48600	2.020	....	....	....	....	....	....	....	....	....	....	....
50000	....	....	....	....	.490	....	....	....	....	....	....	1.160
52000	....	....	....	....	.500	....	....	....	....	....	....	1.230
54000	....	....	....	....	....	....	....	....	....	....	....	1.310
56000	....	....	....	....	....	....	....	....	....	....	....	1.510
57000	....	....	....	....	.540	....	....	....	....	....	....	....
59000	....	....	....	....	....	....	....	....	....	....	....	1.525
61000	....	....	....	....	.630	....	....	....	....	....	....	....
64000	....	....	....	....	.700	....	....	....	....	....	....	....
67000	....	....	....	....	.750	....	....	....	....	....	....	....

Breaking weight of Beam XVII = 48,600 lbs.
" " " XVIII = 69,400 "
" " " XIX = 59,540 "

TABLE G.

Deflections of Beams XX and XXI.

Load in lbs.	XX						XXI				
	1st Loading		Beam gradually relieved of l'd	2nd Loading		1st Loading		Beam gradually relieved of l'd	2nd Loading	
	34½ ins.	End	34½ ins.	Ends.	34½ ins.	Ends	34½ ins.	E'ds	34½ ins.	Ends.	34½ ins.	Ends.	34½ ins.
250	....	....	....	....	....	....	.009	.02	.010	.020	.015	.020	.015
500	.0	.0	.0	.001	.003	.005	.005	.025	.035	.025	....	....	....
750	....	....	....	....	....	....	.0375	.065	.045	....	....	....	....
1000	.015	.016	.015	....	....	....	.055	.085	.060	.095	.065	.095	.065
1250	....	....	....	....	....	....	.075	.110	.075	....	....	....	....
1500	....	....	....	....	....	....	.095	.135	.090	....	....	....	....
1750	....	....	....	....	....	....	.110	.115	.110	....	....	....	....
2000	.035	.040	.040	.045	.035	.045	.040	.120	.185	.125	.195	.120	.185	.125
2250	....	....	....	....	....	....	.140	.205	.140	....	....	....	....
2500	.048	.050	.050	....	....	....	.155	.230	.155	....	....	....	....
3000	.050	.080	.055	....	....	....	.185	.275	.185	.280	.190	.285	.185
3500	.055	.090	.065	....	....	....	.220	.325	.215	....	....	....	....
4000	.070	.105	.080	.105	.070	.105	.075	.255	.370	.250	.375	.255	.370	.250
4250	....	....	....	....	....	....	.270	.405	.270	....	....	....	....
4500	.084	.115	.090	....	....	....	.285	.430	.285	....	....	....	....
4750	....	....	....	....	....	....	.302	.455	.300	....	....	....	....
5000	.095	.125	.100	....	....	....	.317	.470	.315	.470	.315	.470	.315
5250	....	....	....	....	....	....	.335	.495	.330	....	....	....	....
5500	.100	.140	.105	....	....	....	.350	.520	.345	....	....	....	....
5750	....	....	....	....	....	....	.365	.545	.360	....	....	....	....
6000	.110	.155	.115	.160	.105	.160	.110	.380	.565	.375	.575	.380	.575	.380
6500	.120	.170	.120	....	....	....	....	....	....	....	....	....	....
7000	.130	.185	.135	....	....	....	....	....	....	.415	.665	.440
7500	.135	.200	.140	....	....	....	....	....	....	....	....	....
8000	.146	.210	.150	.215	.145	.215	.150	....	....	....	.515	.765	.515
8500	.152	.225	.160	....	....	....	....	....	....	....	....	....
9000	.163	.240	.170	....	....	....	....	....	....	.580	.870	.575
9500	.175	.255	.180	....	....	....	....	....	....	....	....	....
10000	.180	.270	.190	.270	.180	.270	.185	....	....	....	.645	.970	.640
10500	.194	.285	.200	....	....	....	....	....	....	....	....	....
11000	.200	.300	.205	....	....	....	....	....	....	.715	1.075	.700
11500	.210	.315	.220	....	....	....	....	....	....	....	....	....
12000	.220	.325	.230	.325	.215	.325	.235	....	....	....	.785	1.170	.765
13000	....	....	....	....	....	....	....	....	....	....	1.310	....
14000	....	....	....	.255	.380	.260	....	....	....	....	....	....
15000	....	....	....	....	....	....	....	....	....	....	1.515	....
16000	....	....	....	.285	.430	.290	....	....	....	....	1.670	....
17000	....	....	....	....	....	....	....	....	....	....	1.850	....
17400	....	....	....	....	....	....	....	....	....	....	2.000	....
17500	....	....	....	....	....	....	....	....	....	....	2.40	....
18000	....	....	....	.320	.485	.325	....	....	....	....	....	....
20000	....	....	....	.360	.515	.370	....	....	....	....	....	....
22000	....	....	....	.400	.505	.410	....	....	....	....	....	....
24000	....	....	....	.440	.665	.450	....	....	....	....	....	....
26000	....	....	....	.725	....	....	....	....	....	....	....	....
28000	....	....	....	.791	....	....	....	....	....	....	....	....
30000	....	....	....	.850	....	....	....	....	....	....	....	....
32000	....	....	....	.920	....	....	....	....	....	....	....	....
34000	....	....	....	.990	....	....	....	....	....	....	....	....
36000	....	....	....	1.06	....	....	....	....	....	....	....	....
38000	....	....	....	1.50	....	....	....	....	....	....	....	....
40000	....	....	....	2.40	....	....	....	....	....	....	....	....
42000	....	....	....	3.60	....	....	....	....	....	....	....	....
44000	....	....	....	5.05	....	....	....	....	....	....	....	....
46000	....	....	....	6.60	....	....	....	....	....	....	....	....
48000	....	....	....	7.03	....	....	....	....	....	....	....	....

Breaking weight of Beam XX = 49,600 lbs.
" " " XXI = 17,960 "

Tables H and I show deflections in inches of Old Douglas Fir, etc.

TABLE H.

Deflections of Beams XXII and XXIII.

Loads in lbs.	XXII.				XXIII.			
	27 ins.	51 ins.	Ends	54 ins.	27 ins.	31 ins.	62 ins.	Ends.	62 ins.	31 ins.
1,000	.....	.....	.....	.....	.015	.01	.015	.00	.01
1,500	.02	.01	.02	.01	.01	.025	.02	.025	.01	.02
2,000	.025	.02	.03	.01	.02	.04	.03	.045	.02	.035
2,500	.04	.025	.04	.02	.03	.05	.045	.05	.025	.045
3,000	.045	.03	.05	.025	.04	.065	.05	.065	.03	.06
3,500	.05	.035	.06	.03	.05	.08	.06	.085	.04	.07
4,000	.06	.04	.07	.035	.06	.10	.065	.105	.045	.085
4,500	.07	.04	.08	.04	.07	.11	.08	.12	.05	.11
5,000	.08	.05	.10	.045	.08	.125	.09	.135	.06	.115
5,500	.09	.055	.12	.05	.09	.14	.095	.150	.065	.13
6,000	.10	.06	.13	.055	.10	.16	.10	.175	.075	.15
6,500	.11	.06	.14	.055	.11	.17	.11	.185	.075	.16
7,000	.12	.07	.15	.06	.12	.18	.12	.20	.085	.175
7,500	.13	.075	.155	.065	.13	.20	.13	.225	.095	.19
8,000	.14	.08	.16	.07	.14	.21	.14	.25	.10	.20
8,500	.15	.085	.17	.075	.15	.225	.145	.255	.11	.215
9,000	.16	.09	.18	.08	.16	.24	.155	.275	.12	.225
9,500	.17	.095	.195	.085	.17	.25	.160	.285	.125	.245
10,000	.18	.10	.20	.09	.175	.26	.17	.305	.13	.255
10,500	.19	.105	.21	.095	.18	.275	.185	.325	.14	.265
11,000	.195	.11	.22	.10	.19	.29	.19	.345	.145	.275
11,500	.20	.115	.235	.105	.20	.305	.20	.355	.15	.30
12,000	.21	.115	.245	.11	.21	.32	.205	.375	.16	.305
12,500	.22	.12	.255	.115	.22	.335	.21	.390	.17	.32
13,000	.23	.125	.265	.12	.225	.35	.225	.415	.175	.34
13,500	.235	.13	.27	.125	.235	.365	.235	.425	.18	.355
14,000	.25	.14	.29	.13	.25	.38	.245	.44	.19	.365
14,500	.255	.145	.30	.135	.26	.395	.25	.455	.20	.38
15,000	.265	.15	.31	.14	.265	.41	.26	.475	.205	.395
15,500	.27	.155	.32	.145	.27	.425	.27	.495	.215	.405
16,000	.28	.16	.33	.15	.28	.44	.275	.505	.22	.42
16,500	.29	.16	.34	.16	.29	.455	.285	.525	.23	.445
17,000	.295	.17	.35	.165	.29	.47	.29	.545	.245	.45
17,500	.30	.175	.36	.165	.31	.485	.30	.555	.245	.465
18,000	.31	.18	.37	.175	.315	.50	.305	.575	.25	.475
18,500	.32	.185	.39	.175	.32	.515	.313	.595	.26	.485
19,000	.33	.19	.395	.18	.33	.53	.32	.605	.265	.50
19,500	.34	.195	.40	.18	.34	.545	.33	.625	.275	.51
20,000	.35	.20	.42	.185	.35	.555	.345	.645	.28	.53
20,500	.....	.....	.....	.....	.....	.565	.35	.655	.285	.545
21,000	.....	.....	.43	.....	.....	.580	.360	.675	.305	.56
21,500	.....	.....	.....	.....	.....	.59	.37	.695	.305	.57
22,000	.....	.....	.45	.....	.....	.605	.375	.705	.31	.58
22,500	.....	.....	.....	.....	.....	.625	.38	.725	.32	.595
23,000	.....	.....	.....	.....	.....	.645	.395	.745	.325	.61
23,500	.....	.....	.....	.....	.....	.65	.40	.765	.335	.625
24,000	.....	.....	.....	.....	.....	.665	.41	.780	.34	.64
25,000	.....	.....	.51	.....	.....	.....	.....	.....	.....	.....
26,000	.....	.....	.54	.....	.....	.....	.....	.85	.....	.....
27,000	.....	.....	.555	.....	.....	.....	.....	.....	.....	.....
28,000	.....	.....	.57	.....	.....	.....	.....	.90	.....	.....
30,000	.....	.....	.....	.....	.....	.....	.....	1.00	.....	.....
31,000	.....	.....	.66	.....	.....	.....	.....	.....	.....	.....
32,000	.....	.....	.67	.....	.....	.....	.....	1.05	.....	.....
34,000	.....	.....	.71	.....	.....	.....	.....	1.15	.....	.....
35,000	.....	.....	.745	.....	.....	.....	.....	.....	.....	.....
36,000	.....	.....	.76	.....	.....	.....	.....	1.2	.....	.....
38,000	.....	.....	.....	.....	.....	.....	.....	1.27	.....	.....
40,000	.....	.....	.86	.....	.....	.....	.....	1.34	.....	.....
41,000	.....	.....	.90	.....	.....	.....	.....	.....	.....	.....
42,000	.....	.....	.....	.....	.....	.....	.....	1.45	.....	.....
44,000	.....	.....	.975	.....	.....	.....	.....	1.53	.....	.....
45,000	.....	.....	1.02	.....	.....	.....	.....	.....	.....	.....
46,000	.....	.....	.....	.....	.....	.....	.....	1.60	.....	.....
47,000	.....	.....	1.07	.....	.....	.....	.....	.....	.....	.....
49,000	.....	.....	1.10	.....	.....	.....	.....	.....	.....	.....
51,000	.....	.....	1.15	.....	.....	.....	.....	.....	.....	.....
53,000	.....	.....	1.20	.....	.....	.....	.....	.....	.....	.....
55,000	.....	.....	1.27	.....	.....	.....	.....	.....	.....	.....

Breaking weight of Beam XXII = 55,400 lbs.
" " " XXIII = 47,560 "

TABLE I.

Loads in lbs.	Deflections of Beams XXIV and XXV.								
	XXIV.					XXV.			
	22 ins.	41 ins.	Ends	44 ins.	22 ins.	24 ins.	48 ins.	Ends	48 ins.	24 ins.
500	.....	.....	....	.....	.....	.01	.005	.01	.005	.01
1,000	....	.....	....	.....	.....	.015	.01	.015	.005	.015
2,000	....	.....	....	.....	.....	.02	.015	.03	.01	.02
3,000	....	.....	....	.....	.....	.04	.025	.05	.015	.04
4,000	....	.....	....	.....	.....	.06	.035	.075	.025	.06
5,000	.045	.03	.05	.04	.04	.075	.045	.095	.04	.08
6,000	.065	.04	.065	.045	.055	.095	.055	.105	.045	.10
7,000	.08	.04	.08	.05	.06	.115	.065	.140	.055	.115
8,000	.10	.05	.10	.06	.08	.125	.07	.15	.065	.125
9,000	.105	.055	.105	.07	.08	.14	.08	.18	.075	.14
10,000	.12	.06	.12	.07	.095	.155	.09	.195	.08	.155
11,000	.13	.07	.13	.08	.11	.17	.10	.225	.085	.165
12,000	.14	.08	.15	.085	.125	.185	.105	.245	.10	.18
13,000	.115	.085	.16	.09	.14	.205	.115	.26	.105	.21
14,000	.16	.09	.17	.10	.15	.215	.12	.285	.115	.22
15,000	.18	.10	.20	.11	.165	.24	.125	.30	.125	.235
16,000	.20	.105	.21	.12	.17	.255	.14	.325	.13	.255
17,000	.21	.11	.22	.125	.18	.265	.15	.345	.145	.265
18,000	.22	.12	.25	.13	.19	.285	.155	.365	.16	.28
19,000	.225	.125	.25	.14	.205	.30	.16	.395	.17	.305
20,000	.24	.13	.26	.15	.22	.315	.17	.410	.18	.315
21,000	.26	.14	.27	.16	.24	.340	.185	.445	.19	.335
22,000	.27	.145	.29	.17	.25	.355	.195	.465	.20	.355
23,000	.28	.15	.31	.175	.26	.....	.....	.....	....	....
24,000	.30	.16	.32	.18	.27	.....	.....	.50	....	....
25,000	.31	.17	.335	.185	.275	.....	.....	.....	....	....
25,800	....	....	....	....	....	.....	.....	.54	....	....
26,000	.32	.175	.35	.195	.29	.....	.....	.....	....	....
27,000	.34	.18	.36	.205	.31	.....	.....	.....	....	....
28,000	.36	.18	.38	.21	.32	.....	.....	.....	....	....
29,000	.37	.19	.40	.22	.33	.....	.....	.....	....	....
30,000	.38	.20	.415	.225	.34	.....	.....	.....	....	....
30,200	....	....	....	....	....	.....	.....	.65	....	....
31,000	.39	.21	.425	.235	.355	.....	.....	.....	....	....
32,000	.405	.22	.45	.24	.37	.....	.....	.....	....	....
33,000	....	....	.46	....	....	.....	.....	.....	....	....
33,200	....	....	....	....	....	.....	.....	.75	....	....
34,000	....	....	.48	....	....	.....	.....	.....	....	....
36,000	....	....	.51	....	....	.....	.....	.....	....	....
37,000	....	....	.54	....	....	.....	.....	.....	....	....
38,000	....	....	.56	....	....	.....	.....	.....	....	....
39,000	....	....	.575	....	....	.....	.....	.....	....	....
39,700	....	....	....	....	....	.....	.....	.95	....	....
40,000	....	....	.66	....	....	.....	.....	.....	....	....

Breaking weight of Beam XXIV = 76,900 lbs. for beam of reduced length.
Breaking weight of Beam XXV = 42,900 lbs.

Table J showing deflections in inches of two Douglas Fir planks under gradually increased loads.

Loads in lbs.	Deflections in ins. of Plank 1.	Deflections in ins. of Plank 2.
	Ends.	Ends.
2,000	.05	.06
3,000	.07	.10
4,000	.10	.15
5,000	.12	.19
6,000	.15	.23
7,000	.16	.27
8,000	.18	.35
9,000	.21	...

Breaking weight of Plank 1 = 22,250 lbs.
" " " 2 = 13,250 "

Tables K to M show deflections in inches of Canadian New Red Pine Beams.

TABLE K.

Loads in lbs.	Deflections of Beams XXVI to XXVIII.						
	XXVI.					XXVII	XXVIII.
	35 ins.	70 ins.	Ends.	70 ins.	35 ins.	Ends.	Ends.
1,000	.055	.035	.065	.04	.055	.08	.09
1,500	.110	.060	.135	.060	.110	.15	.15
1,800	.145	.080	.175	.080	.150	......	......
2,000	.165	.095	.200	.09	.165	.20	.225
2,300	.195	.110	.235	.110	.200	......	......
2,500	.215	.125	.260	.125	.215	.26	.306
2,700	.235	.130	.285	.130	.240	......	......
3,000	.265	.150	.320	.150	.265	.32	.36
3,200	.290	.160	.350	.160	.295	......	......
3,500	.320	.180	.385	.180	.320	.37	.44
3,700	.345	.195	.410	.195	.350	......	......
4,000	.370	.210	.450	.210	.370	.44	.50
4,200	.395	.225	.475	.225	.400	......	......
4,500	.430	.245	.510	.245	.430	.49	.575
4,700	.450	.255	.535	.250	.450	......	......
5,000	.480	.270	.570	.265	.475	.55	.65
5,200	.500	.280	.600	.275	.500	......	......
5,500	.535	.295	.635	.290	.530	.60	.72
5,700	.560	.310	.660	.305	.550	......	......
6,000	.580	.330	.700	.320	.580	.66	.79
6,200	.605	.340	.725	.335	.600	......	......
6,500	.635	.360	.755	.350	.635	.73	.86
6,700	.655	.370	.790	.365	.655	......	......
7,000	.690	.385	.825	.380	.685	.79	.93
7,200	.715	.395	.855	.390	.705	......	......
7,500	.745	.415	.890	.410	.740	.85	1.00
7,700	.765	.425	.915	.425	.755	......	......
8,000	.800	.445	.950	.440	.800	.92	1.07
8,200	.820	.455	.980	.455	.815	......	......
8,500	.850	.475	1.020	.470	.855	.99	1.14
8,700	.880	.495	1.050	.485	.875	......	......
9,000	.915	.510	1.100	.510	.915	1.05	1.21
9,200	.945	.525	1.135	.525	.945	......	......
9,500	.995	.545	1.185	.545	.985	1.13	1.28
9,700	1.015	.560	1.225	.560	1.010	......	......
10,000	1.050	.585	1.265	.580	1.050	1.20	1.36
10,500	......	......	......	......	......	......	1.43
11,000	......	......	1.400	......	......	1.36	1.50
11,500	......	......	......	......	......	......	1.57
12,000	......	......	1.600	......	......	1.51	1.66
12,500	......	......	......	......	......	......	1.72
13,000	......	......	1.700	......	......	1.63	1.80
13,500	......	......	......	......	......	......	1.87
13,800	......	......	......	......	......	......	......
14,000	......	......	2.050	......	......	......	1.95
14,500	......	......	......	......	......	......	2.06
15,000	......	......	......	......	......	2.00	2.15
15,500	......	......	......	......	......	......	2.30
15,600	......	......	2.750	......	......	......	......
16,000	......	......	3.000	......	......	2.20	2.44
16,500	......	......	......	......	......	......	......
17,000	......	......	......	......	......	2.52	......
17,050	......	......	......	......	......	......	2.80

Breaking weight of Beam XXVI = 16,940 lbs.
" " " XXVII = 17,700 "
" " " XXVIII = 17,050 "

TABLE L.

Loads in lbs.	Deflections of Beams XXIX to XXXII.							
	XXIX.					XXX.	XXXI.	XXXII.
	35 ins.	70 ins.	Ends.	70 ins.	35 ins.	Ends.	Ends.	Ends.
200	......	....	....	....	....	....	....	.035
500	.030	.015	.04	.015	.020	.130	....	.185
600	......	....	....	....	....	....	....	.235
700	......	....	....	....	....	....	....	.290
800	......	....	....	....	....	.245	....	.340
900	......	....	....	....	....	....	....	.385
1,000	.120	.050	.140	.070	.100	.320	.29	.430
1,100	......	....	....	....	....	....	....	.495
1,200	......	....	....	....	....	....	....	.545
1,300	......	....	....	....	....	.410	.385	.600
1,400	.185	.090	.225	.110	.190	....	....	.650
1,500	......	....	......	....	....	.505	.450	.700
1,600	......	....	....	....	....	....	....	.750
1,700	......	....	....	....	....	.590	.520	.800
1,800	.265	.135	.310	.150	.250	....	....	.855
1,900	......	....	....	....	....	....	....	.915
2,000	.300	.150	.350	.170	.290	.710	.615	.960
2,100	......	....	....	....	....	....	....	1.015
2,200	......	....	....	....	....	....	....	1.075
2,300	......	....	....	....	....	.835	.725	1.145
2,400	.370	.190	.440	.205	.360	....	....	1.195
2,500	......	....	....	....	....	.905	.780	1.245
2,600	......	....	....	....	....	....	....	1.300
2,700	......	....	....	....	....	....	....	1.360
2,800	.440	.235	.525	.230	.435	1.040	.900	1.410
2,900	......	....	....	....	....	....	....	1.465
3,000	.480	.250	.565	.265	.460	1.150	.960	1.525
3,100	......	....	....	....	....	....	....	1.585
3,200	......	....	....	....	....	1.210	1.035	1.625
3,300	......	....	....	....	....	....	....	1.700
3,400	.550	.295	.650	.305	.540	....	....	1.750
3,500	......	....	....	....	....	1.310	1.115	1.800
3,600	......	....	....	....	....	....	....	1.865
3,700	......	....	....	....	....	....	....	1.935
3,800	.620	.330	.740	.350	.610	1.455	1.225	1.990
3,900	......	....	....	....	....	....	....	2.025
4,000	.640	.350	.775	.365	.640	1.550	1.320	2.100
4,100	......	....	....	....	....	....	....	2.170
4,200	......	....	....	....	....	1.640	....	2.220
4,300	......	....	....	....	....	....	1.415	2.290
4,400	.740	.390	.865	.410	.730	....	....	2.355
4,500	......	....	....	....	....	1.765	1.510	2.420
4,600	......	....	....	....	....	....	....	2.470
4,700	......	....	....	....	....	....	....	2.530
4,800	.810	.415	.960	.450	.800	1.900	1.615	2.610
4,900	......	....	....	....	....	....	....	2.680
5,000	.850	.460	1.000	.470	.835	2.010	1.700	2.755
5,100	......	....	....	....	....	....	....	2.830
5,200	......	....	......	....	....	2.120	....	......
5,300	......	....	....	....	....	....	1.815	......
5,400	.910	.500	1.085	.515	.900	....	....	......
5,500	......	....	....	....	....	2.335	1.895	......
5,700	....	....	....	....	....	2.515	....	......
5,800	.985	.545	1.175	.560	.990	....	....	......
6,000	1.030	.565	1.225	.580	1.005	2.900	2.115	......
6,400	1.110	.610	1.320	.620	1.100	....	....	......
6,500	......	....	....	....	....	....	2.410	......
6,800	1.170	.640	1.405	.660	1.175	....	....	......
7,000	1.220	.665	1.455	.675	1.210	....	....	......
7,400	1.290	.715	1.555	.740	1.300	....	....	......
7,800	1.360	.755	1.660	.775	1.360	......	....	......
8,000	1.410	.785	1.710	.800	1.410	....	....	......
8,400	1.500	.830	1.810	.850	1.510	....	....	......
8,800	1.590	.880	1.915	.900	1.580	....	....	......
9,000	1.640	.910	2.005	.930	1.650	....	....	......
10,000	......	....	2.270	....	....	....	....	......
11,000	......	....	2.650	....	....	....	....	......

Breaking weight of Beam XXIX = 11,960 lbs.
" " " XXX = 5,700 "
" " " XXXI = 6,500 "
" " " XXXII = 5,200 "

TABLE M.

Deflections of Beams XXXIII to XXXV.

Loads in lbs.	XXXIII. Ends.	XXXIV. Ends.	XXXV. Ends.
500	.065	.080	.030
800	.....	.145	.065
1,000	.160	.185	.090
1,200	.205	.230	.125
1,400	.250	.275	.150
1,600	.275	.320	.175
1,800	.325	.360	.195
2,000	.375	.405	.220
2,200	.410	.450	.245
2,400	.465	.490	.270
2,600	.500	.535	.295
2,800	.540	.580	.320
3,000	.585	.625	.345
3,200	.630	.670	.370
3,400	.670	.715	.390
3,600	.710	.760	.415
3,800	.750	.810	.445
4,000	.790	.850	.465
4,200	.830	.900	.490
4,400	.870	.945	.515
4,600	.910	.990	.545
4,800	.950	1.035	.565
5,000	1.000	1.080	.590
5,200	1.040	1.125	.615
5,400	1.090	1.175	.640
5,600	1.125	1.220	.670
5,800	1.165	......	.695
6,000	1.220	......	.720
6,200	1.260	......	.745
6,400	1.310	......	.770
6,600	1.355	......	.800
6,800	1.415	......	.830
7,000	1.455	......	.860
7,200	1.545	......	.885
7,400	1.590	......	.915
7,600	1.640	......	.950
7,800	1.690	......	......
8,200	1.790	......	......

Breaking weight of Beam XXXIII = 9,250 lbs.
    "   "   "   XXXIV = 5,600 "
    "   "   "   XXXV = 7,600 "

Tables N to Q show deflections in inches of Canadian New White Pine Beams.

TABLE N.

Deflections of Beams XXXVI to XLI.

Loads in lbs.	XXXVI. 108 ins.	72 ins.	36 ins.	Ends.	36 ins.	72 ins.	108 ins.	XXXVII. Ends.	XXXVIII. Ends.	XXXIX. Ends.	XL. Ends.	XLI. Ends.
5000	.109	.30	.30	.32	.30	.29	.109	......	......	......	....	......
7500	.375	.70	.93	1.02	.90	.66	.344	......	...	......	....	......
10000	.594	1.00	1.83	1.45	1.29	.95	.516	......	.10	.11	.11	.13
11000	.719	1.34	1.78	1.95	1.74	1.28	.688	......	......	......	....	......
12500	.799	1.47	1.96	2.16	1.93	1.42	.750	......	.125	.14	....	......
15000	.906	1.68	2.24	2.45	2.20	1.62	.875	......	.15	.165	.17	.20
17500	1.125	2.05	2.70	2.97	2.65	1.96	1.047	......	.19	.19	....	......
20000	....	....	....	....	....	....	....	......	.21	.2255	.23	.29
22000	....	....	....	....	....	....	......	......	......	......	.25	.32
22500	......	......	....	....	....	....	......	......	.245	.2555	....	......
24000	....	....	....	....	......	....	....	......	......	......	.27	.35
25000	....	....	....	....	....	....	....	......	.27	.285	....	......
26000	....	....	....	......	....	....	....	......	......	......	.30	.40
27500	....	....	....	....	....	....	....	......	.30	.31	....	......
28000	....	....	....	......	......	....	....	......	......	......	.33	.44
30000	....	....	....	....	....	......	....	......	.33	.35	.36	.49
32000	....	....	....	....	......	......	....	......	......	......	.39	.53
32500	....	....	....	....	....	......	....	......	.37	......	....	......
34000	......	......	....	....	......	......	....	......	......	......	.42	......
36000	......	......	....	....	....	......	....	......	......	......	.45	......

Breaking weight of Beam XXXIV = 19,600 lbs.
    "   "   "   XXXV = 24,000 "
    "   "   "   XXXVI = 52,450 "
    "   "   "   XXXVII = 51,400 "

TABLE O.

| Loads in lbs. | Deflections of Beams XL to XLII. |||||| XLIII. | XLIV. |
| | XLII. |||||| | |
	108 ins	72 ins.	36 ins.	Ends.	36 ins.	72 ins.	108 ins	Ends.	Ends.
2500	.0312	.05	.07	.08	.07	.055	.031	......	......
3000	.047	.095	.14	.15	.14	.10	.047	......	......
3500	.078	.13	.18	.19	.18	.13	.078	......	......
4000	.094	.17	.24	.26	.24	.17	.109	......	......
4500	.109	.20	.27	.30	.28	.205	.125	......	......
5000	.125	.245	.33	.37	.34	.25	.141	......	......
5500	.141	.275	.38	.42	.39	.28	.156	......	......
6000	.172	.325	.44	.47	.45	.33	.172	......	......
6500	.187	.35	.49	.53	.49	.35	.188	......	......
7000	.219	.39	.54	.60	.54	.40	.219	......	......
7500	.234	.425	.59	.65	.60	.43	.234	......	......
8000	.250	.47	.64	.71	.65	.47	.266	......	......
8500	.281	.505	.69	.76	.70	.52	.281	......	......
9000	.297	.54	.75	.82	.75	.55	.312	......	......
9500	.312	.59	.80	.90	.81	.60	.328	......	......
10000	.328	.61	.84	.93	.85	.63	.344	.10	.11
10500	.359	.66	.91	1.00	.91	.67	.359	......	......
11000	.375	.70	.97	1.07	.96	.71	.375	......	......
11500	.406	.75	1.03	1.14	1.04	.76	.406	......	......
12000	.422	.77	1.06	1.17	1.07	.79	.422	......	......
12500	.438	.80	1.11	1.21	1.11	.82	.438	......	......
13000	.453	.835	1.16	1.30	1.17	.875	.453	......	......
13500	.484	.905	1.24	1.37	1.25	.93	.484	......	......
14000	.500	.945	1.29	1.44	1.31	.97	.510	......	......
14500	.531	.975	1.34	1.49	1.355	1.00	.531	......	......
15000	.547	1.02	1.40	1.55	1.415	1.02	.562	.16	.16
15500	.562	1.06	1.45	1.61	1.48	1.10	.578	......	....
16000	.593	1.105	1.51	1.68	1.53	1.15	.593	......	....
16500	.609	1.15	1.57	1.76	1.60	1.19	.625	......	....
17000	.641	1.19	1.63	1.81	1.65	1.23	.641	......	......
17500	.656	1.23	1.68	1.87	1.705	1.27	.672	......	......
18000	.687	1.27	1.75	1.96	1.775	1.32	.687	......	......
18500	.719	1.34	1.84	2.05	1.86	1.39	.734	......	......
19000	.750	1.38	1.89	2.11	1.92	1.43	.750	......	......
19500	.766	1.43	1.95	2.19	1.98	1.47	.766	......	......
20000	.781	1.48	2.02	2.27	2.05	1.52	.797	.23	.24
20500	.813	1.53	2.10	2.35	2.13	1.58	.828	......	....
21000	.844	1.58	2.16	2.42	2.19	1.62	.859	......	....
21500	.875	1.665	2.28	2.55	2.31	1.70	.891	......	......
22000	.924	1.72	2.36	2.65	2.39	1.77	.938	....	......
25000	....	....	....	....	......	....	.....	.29	.30

Breaking weight of Beam XXXVIII = 26,350 lbs.
" " " XXXIX = 48,600 "
" " " XL = 51,870 "

TABLE P.

Deflections of Beams XLV to XLVII.

Loads in lbs.	XLV.						XLVI.	XLVII.
	108 ins	72 ins.	36 ins	Ends, 36 ins.	72 ins.	108 ins	Ends.	Ends.
2500	.125	.22	.30	.34	.29	.21	......	.02
3000	.141	.27	.35	.39	.34	.31	......	......
3500	.172	.29	.41	.45	.39	.34	.188	......
4000	.188	.34	.45	.50	.41	.36	.203	......
4500	.203	.37	.50	.55	.49	.44	.219	......
5000	.219	.42	.55	.61	.54	.44	.234	......
5500	.234	.45	.60	.67	.59	.47	.250	......
6000	.250	.49	.65	.73	.64	.51	.266	......
6500	.266	.53	.71	.79	.69	.56	.281	......
7000	.297	.56	.76	.84	.74	.59	.312	......
7500	.312	.60	.81	.90	.79	.62	.328	......
8000	.344	.63	.86	.95	.85	.66	.344	......
8500	.359	.67	.92	1.03	.90	.69	.359	......
9000	.375	.71	.97	1.08	.95	.74	.391	......
9500	.391	.75	1.02	1.14	1.00	.78	.406	......
10000	.422	.79	1.08	1.20	1.06	.81	.422	.10
10500	.438	.83	1.14	1.26	1.11	.86	.438	......
11000	.453	.87	1.20	1.33	1.17	.90	......	......
11500	.484	.92	1.26	1.40	1.24	.95	.500	......
12000	.500	.96	1.31	1.47	1.28	.98	.516	......
12500	.531	1.01	1.36	1.53	1.34	1.02	.531	.13
13000	.547	1.05	1.42	1.59	1.39	1.06	.547	......
13500	.563	1.08	1.48	1.66	1.45	1.10	.578	......
14000	.593	1.13	1.55	1.73	1.51	1.15	.593	......
14500	.625	1.17	1.60	1.79	1.57	1.18	.625	......
15000	.641	1.21	1.65	1.86	1.62	1.22	.641	.20 · .16
15500	.656	1.25	1.71	1.93	1.69	1.27	.656	......
16000	.687	1.30	1.78	2.00	1.75	1.31	.672	......
16500	.703	1.35	1.85	2.08	1.82	1.36	.687	......
17000	.734	1.39	1.90	2.14	1.86	1.40	.734	......
17500	.766	1.43	1.97	2.22	1.94	1.45	.750	.20
18000	.781	1.50	2.05	2.33	2.02	1.51	.781	......
18500	.797	1.54	2.11	2.39	2.08	1.56	.797	......
19000	.828	1.59	2.19	2.48	2.15	1.60	.828	......
20000	.875	1.68	2.31	2.63	2.29	1.70	.875	.26 · .23
20500	.924	1.75	2.41	2.76	2.38	1.77	.924	......
21000	.953	1.82	2.50	2.88	2.47	1.83	.953	......
22500	......	......	......	......	......	......	......	.26
25000	......	......	......	......	......	......	.35	.30
27500	......	......	......	......	......	......	......	.34
30000	......	......	......	......	......	......	......	.39

Breaking weight of Beam XLI = 24,850 lbs.
" " " XLII = 41,400 "
" " " XLIII = 48,650 "

TABLE Q.

Loads in lbs.	Deflections of Beams XLVIII to L.								
	XLVIII.			XLIX.			L.		
	37½ ins.	Ends.	37½ ins.	37½ ins.	Ends.	37½ ins.	46½ ins.	Ends.	46½ ins.
1000	.01	.01	.01	.005	.01	.005	.015	.015	.01
2000	.025	.03	.02	.02	.04	.02	.04	.055	.035
3000	.04	.05	.035	.035	.06	.035	.07	.105	.065
4000	.055	.065	.052	.05	.08	.05	.10	.15	.10
5000	.065	.085	.06	.065	.10	.065	.135	.195	.135
6000	.08	.105	.075	.075	.125	.08	.165	.245	.165
7000	.10	.125	.08	.095	.15	.095	.20	.295	.20
8000	.105	.15	.103	.11	.17	.105	.22	.33	.225
9000	.12	.17	.11	.125	.20	.13	.25	.375	.255
10000	.135	.195	.125	.14	.22	.14	.28	.43	.28
10500	.14	.215	.135	....	....	....	......	......	......
11000	.15	.22	.143	.155	.25	.15	.30	.46	.30
11500	.155	.23	.15	....	....	....	......	......	......
12000	.165	.24	.155	.175	.265	.165	.33	.50	.33
12500	.175	.25	.16	.18	.275	.17	.35	.53	.35
13000	.18	.265	.165	.19	.29	.185	.36	.55	.36
13500	.185	.27	.17	.20	.30	.195	.375	.57	.375
14000	.19	.285	.177	.21	.315	.20	.39	.60	.39
14500	.20	.295	.19	.215	.32	.21	.41	.615	.40
15000	.21	.305	.20	.22	.35	.215	.42	.645	.42
15500	.215	.32	.205	.225	.355	.22	.43	.655	.43
16000	.22	.33	.21	.235	.365	.23	.445	.67	.45
16500	.23	.34	.223	.245	.375	.24	.46	.70	.46
17000	.235	.355	.23	.25	.39	.25	.475	.72	.475
17500	.24	.365	.235	.26	.405	.255	.49	.745	.50
18000	.25	.38	.24	.27	.415	.26	.51	.76	.51
18500	.25	.395	.25	.275	.425	.27	.525	.795	.52
19000	.265	.405	.255	.285	.44	.28	.54	.82	.55
19500	.27	.415	.26	.295	.455	.29	.55	.84	.56
20000	.275	.425	.27	.30	.465	.30	.57	.865	.58
20500	.285	.445	.285	.31	.475	.31	.585	.895	.59
21000	.295	.46	.29	.32	.495	.32	.60	.92	.61
21500	.30	.47	.295	.325	.505	.325	.62	.94	.63
22000	.31	.485	.305	.34	.515	.335	.635	.965	.64
22500	.32	.50	.31	.345	.52	.34	.65	1.00	.65
23000	.33	.515	.32	.35	.535	.345	......	1.03	......
23500	.335	.53	.33	.36	.555	.35	......	......	......
24000	.35	.54	.34	.37	.57	.36	......	1.07	......
24500	.36	.555	.35	.38	.58	.37	......	......	......
25000	.365	.565	.355	.385	.585	.375	......	1.14	......
25500	.375	.585	.365	.39	.60	.385	......	......	......
26000	.385	.60	.38	.40	.61	.395	......	1.16	......
26500	.395	.615	.385	.415	.625	.405	......	......	......
27000	....	.625	....	.42	.643	.41	......	1.25	......
27500	....	....	....	.43	.66	.42	......	......	......
28000	....	....	....	.445	.675	.43	......	1.33	......
28500	....	....	....	.45	.69	.445	......	......	......
29000	....	....	....	.46	.71	.455	......	1.41	......
29500	....	....	....	.465	.725	.46	......	......	......
30000	....	.69	....	.475	.74	.47	......	1.49	......
31000	....	....	......	....	.78	....	......	1.55	......
32000	....	.76	......	....	....	....	......	1.60	......
34000	....	.85	....	....	....	....	......	......	......
36000	....	.94	....	....	.92	....	......	......	......
37000	....	....	....	....	.98	....	......	......	......
37300	....	....	....	....	1.00	....	......	......	......
38100	....	1.18	....	....	....	....	......	......	......
40000	....	1.25	......	....	1.20	....	......	......	......
41000	....	......	....	....	1.30	....	......	......	......
44000	....	....	....	....	1.50	....	......	......	......
45000	....	1.85	....	....	....	....	......	......	......
46000	....	1.97	....	....	1.70	....	......	......	......
47000	....	2.15	....	....	1.95	....	......	......	......

Breaking weight of Beam XLVIII = 38,100 lbs.
" " " XLIX = 47,080 "
" " " L = 32,200 "

Table R shows deflections in inches of Canadian White Pine Beams which have been in service.

TABLE R.

Deflections of Beams LI to LIII.

Loads in lbs.	LI.				LII.						LIII.		
	32 ins.	64 ins.	Ends.	64 ins.	32 ins.	30 ins.	60 ins.	Ends.	60 ins.	30 ins.	30 ins.	60 ins.	Ends.	60 ins.	30 ins.
1000	.02	.02	.035	.02	.02	.02	.01	.025	.01	.02	.03	.01	.04	.02	.03
1500	.05	.03	.065	.03	.05	.05	.02	.055	.025	.05	.055	.02	.065	.04	.06
2000	.06	.05	.09	.05	.07	.060	.040	.075	.040	.070	.08	.04	.10	.05	.085
2500	.10	.065	.12	.06	.10	.09	.05	.105	.05	.095	.11	.06	.125	.065	.11
3000	.11	.08	.145	.07	.12	....	....	....	....	....	.135	.08	.16	.08	.14
3200	....	....	....	....	....	.12	.06	.135	.07	.125	....	....	....	....	....
3500	.135	.09	.175	.085	.15	.14	.07	.155	.08	.145	.16	.095	.20	.09	.16
4000	.17	.10	.21	.10	.175	.16	.08	.185	.09	.16	.18	.105	.235	.10	.19
4500	.19	.12	.24	.115	.20	.18	.10	.21	.11	.18	.21	.11	.26	.12	.22
5000	.21	.13	.265	.13	.23	.20	.105	.235	.12	.205	.235	.13	.28	.13	.24
5500	.25	.14	.30	.145	.25	....	....	....	....	....	.26	.145	.325	.15	.27
5700	....	....	....	....	....	.22	.12	.265	.13	.245	....	....	....	....	....
6000	.27	.15	.325	.16	.275	.245	.13	.285	.14	.25	.29	.16	.35	.165	.30
6500	.29	.17	.35	.17	.30	.26	.14	.31	.155	.275	.31	.18	.39	.18	.32
7000	.31	.185	.385	.185	.33	.29	.15	.345	.175	.30	.34	.19	.42	.19	.35
7500	.345	.20	.415	.20	.35	....	....	....	....	....	.37	.20	.45	.21	.385
7800	....	....	....	....	....	.31	.16	.375	.19	.325	....	....	....	....	....
8000	.35	.21	.445	.215	.375	.34	.17	.40	.20	.35	.40	.22	.49	.23	.405
8500	.38	.225	.47	.235	.40	.35	.185	.415	.215	.36	.425	.24	.515	.24	.44
9000	.40	.23	.50	.25	.425	.375	.195	.445	.225	.39	.455	.25	.55	.255	.46
9500	.425	.25	.53	.26	.45	.40	.21	.475	.24	.41	.47	.27	.585	.27	.495
10000	.45	.26	.555	.285	.48	.42	.22	.50	.25	.435	.505	.285	.615	.285	.52
10500	.47	.27	.585	.29	.50	.45	.24	.535	.27	.46	.53	.29	.65	.30	.55
11000	.50	.29	.615	.305	.53	....	....	....	....	....	.565	.305	.69	.31	.58
11500	.515	.30	.65	.315	.55	.47	.25	.56	.28	.485	.59	.32	.725	.33	.60
12000	.55	.31	.67	.33	.58	....	....	....	....	.625	.34	.76	.35	.64
12500	.57	.33	.70	.35	.60	.51	.27	.615	.31	.53	.65	.355	.795	.365	.665
13000	.60	.34	.735	.36	.63	.55	.30	.655	.33	.57	.675	.365	.825	.39	.69
13500	.62	.35	.76	.37	.66	.57	.31	.685	.315	.59	.71	.385	.855	.405	.72
14000	.65	.365	.79	.39	.685	.60	.32	.71	.355	.61	.74	.405	.90	.42	.75
14500	.67	.38	.82	.40	.71	.615	.34	.74	.37	.64	.77	.42	.94	.43	.79
15000	.70	.39	.85	.415	.735	.64	.35	.765	.385	.655	.80	.435	.985	.45	.815
15500	.725	.41	.875	.435	.76	.66	.36	.79	.39	.68	.835	.46	1.02	.47	.85
16000	.75	.42	.91	.446	.785	.69	.38	.83	.415	.71	.87	.47	1.07	.48	.89
16500	.77	.435	.94	.455	.81	....	....	....	....	....	....	....	....	....	....
17000	.80	.45	.97	.47	.84	.72	.395	.865	.43	.74	....	....	1.15	....	....
17500	.82	.47	1.00	.49	.86	.76	.415	.915	.45	.78	....	....	....	....	....
18000	.85	.475	1.03	.51	.89	.79	.44	.95	.47	.81	....	....	....	....	....
18500	.88	.49	1.07	.53	.925	....	....	....	....	....	....	....	....	....	....
19000	.90	.50	1.10	.54	.96	....	....	.985	....	....	....	....	....	....	....
19500	.93	.52	1.14	.56	.985	....	....	....	....	....	....	....	....	....	....
20000	.96	.54	1.185	.60	1.03	....	....	1.06	....	....	....	....	....	....	....
20500	1.00	....	1.235	....	1.07	....	....	....	....	....	....	....	....	....	....
21000	1.04	....	1.28	....	1.11	....	....	1.10	....	....	....	....	....	....	....
21500	....	....	1.32	....	....	....	....	....	....	....	....	....	....	....	....
22000	....	....	....	....	....	....	....	1.18	....	....	....	....	....	....	....
22650	....	....	1.40	....	....	....	....	....	....	....	....	....	....	....	....
23500	....	....	....	....	....	....	....	1.30	....	....	....	....	....	....	....
24000	....	....	....	....	....	....	....	1.34	....	....	....	....	....	....	....
25000	....	....	....	....	....	....	....	1.46	....	....	....	....	....	....	....

Breaking weight of Beams LI = 22,730 lbs.
" " " LII = 26,320 "
" " " LIII = 18,000 "

Tables S and T show deflections in inches of Canadian New Spruce Beams (B.C.).

TABLE S.

Loads in lbs.	Deflections of Beam LIV.						
	108 ins.	72 ins.	36 ins.	Ends.	36 ins.	72 ins.	108 ins.
1,000	.14	.22	.30	.30	.26	.20	.11
1,500	.15	.24	.33	.34	.30	.23	.12
2,000	.17	.28	.37	.38	.34	.25	.15
2,500	.18	.31	.41	.43	.38	.29	.16
3,000	.19	.34	.44	.46	.42	.31	.18
3,500	.21	.36	.48	.51	.45	.34	.19
4,000	.22	.39	.52	.56	.50	.37	.21
4,500	.24	.42	.56	.60	.54	.39	.22
5,000	.25	.45	.60	.61	.57	.42	.24
5,500	.26	.47	.63	.68	.60	.45	.25
6,000	.27	.50	.67	.72	.64	.48	.26
6,500	.29	.53	.71	.76	.67	.50	.28
7,000	.31	.56	.75	.80	.71	.52	.30
7,500	.32	.59	.79	.84	.75	.56	.31
8,000	.34	.61	.82	.88	.79	.60	.32
8,500	.35	.65	.86	.92	.83	.61	.34
9,000	.37	.67	.90	.97	.86	.65	.35
9,500	.38	.70	.94	1.01	.90	.67	.36
10,000	.40	.73	.97	1.05	.94	....	.39
10,500	.41	.76	1.01	1.09	.98	.71	.40
11,000	.43	.79	1.05	1.14	1.02	.72	.41
11,500	.44	.84	1.09	1.17	1.05	.75	.43
12,000	.46	.84	1.13	1.21	1.09	.78	.45
12,500	.48	.87	1.16	1.26	1.14	.82	.46
13,000	.49	.89	1.19	1.29	1.16	.83	.48
13,500	.50	.92	1.23	1.31	1.20	.84	.49
14,000	.51	.95	1.27	1.38	1.24	....	.50
14,500	.53	.98	1.30	1.42	1.28	....	.51
15,000	.54	.99	1.32	1.45	1.31	....	.53
15,500	.55	1.00	1.32	1.46	1.32	.99	.54
16,000	.55	1.00	1.33	1.48	1.34	1.01	.54
16,500	.55	1.01	1.34	1.50	1.35	1.02	.55
17,000	.56	1.01	1.34	1.51	1.36	1.03	.56
17,500	.56	1.02	1.35	1.52	1.40	1.05	.57
18,000	.56	1.03	1.35	1.54	1.41	1.06	.58
18,500	.57	1.03	1.36	1.55	1.43	1.07	.59
19,000	.57	1.04	1.36	1.57	1.45	1.09	.60
19,500	.58	1.04	1.36	1.58	1.46	1.11	.60
20,000	.58	1.05	1.37	1.60	1.47	1.12	.61
20,500	.71	1.32	1.52	1.93	1.74	1.30	.70
21,000	.72	1.35	1.80	1.98	1.78	1.33	.71
21,500	.74	1.38	1.85	2.02	1.82	1.36	.73
22,000	.76	1.41	1.90	2.07	1.86	1.39	.75
23,400	....	....	....	2.20	....	....	....
26,200	....	....	....	2.50	....	....	....
27,600	....	....	....	2.75	....	....	....
29,000	....	....	....	2.85	....	....	....
29,900	....	....	....	3.00	....	....	....
30,800	....	....	....	3.15	....	....	....
32,000	....	....	....	3.25	....	....	....
32,500	....	....	....	3.35	....	....	....
33,200	....	....	....	3.70	....	....	....
33,500	....	....	....	3.80	....	....	....
33,800	....	....	....	4.00	....	....	....
34,400	....	....	....	4.10	....	....	....
34,800	....	....	....	4.25	....	....	....
35,600	....	....	..	4.50	....	....	....
36,200	....	....	....	4.60	....	....	....
36,300	....	....	....	4.75	....	....	....
36,600	....	....	....	4.90	....	....	....
36,800	....	....	....	5.00	....	....	....
38,250	....	....	....	5.50	....	....	....

Breaking weight of Beam LIV = 36,800 lbs.

TABLE T.

Deflections of Beams LV and LVI.

Loads in lbs.	LV			LVI		
	30 ins.	End.	30 ins.	30 ins.	End.	30 ins.
10,000	.05	.09	.05	.1	.07	.0
11,000	.06	.10	.06	.11	.09	.06
12,000	.07	.10	.065	.12	.10	.06
13,000	.07	.11	.07	.13	.10	.07
14,000	.08	.11	.075	.13	.11	.08
15,000	.08	.12	.08	.135	.12	.09
16,000	.09	.13	.085	.14	.13	.09
17,000	.10	.14	.09	.145	.14	.095
18,000	.10	.15	.095	.15	.15	.10
19,000	.11	.16	.105	.16	.15	.105
20,000	.11	.17	.11	.16	.16	.11
21,000	.12	.17	.12	.17	.17	.115
22,000	.12	.18	.125	.175	.18	.12
23,000	.13	.19	.13	.185	.19	.12
24,000	.13	.20	.135	.19	.19	.13
25,000	.14	.21	.14	.195	.20	.14
26,000	.15	.22	.145	.2	.20	.15
27,000	.15	.23	.15	.2	.22	.15
28,000	.16	.24	.16	.215	.24	.16
29,000	.16	.25	.165	.22	.24	.16
30,000	.17	.26	.17	.225	.25	.17
31,000	.17	.27	.18	.23	.26	.17
32,000	.18	.28	.185	.235	.27	.18
33,000	.19	.29	.19	.24	.28	.185
34,000	.20	.30	.20	.245	.29	.19
35,000	.20	.31	.205	.255	.29	.20
36,000	.21	.32	.21	.267	.31	.20
37,000	.21	.33	.215	.27	.32	.21
38,000	.22	.34	.225	.28	.33	.215
39,000	.22	.35	.23	.28	.34	.225
40,000	.23	.36	.24	.285	.35	.235
41,000	.24	.37	.25	.29	.36	.24
42,000	.25	.38	.255	.30	.37	.25
43,000	.25	.39	.26	.31	.39	.255
44,000	.26	.40	.27	.32	.40	.26
45,000	.27	.41	.28	.325	.41	.27
46,000	.27	.42	.29	.335	.42	.28
47,000	.28	.41	.30	.34	.45	.285
48,000	.29	.45	.305	.35	.46	.30
49,000	.30	.46	.315	.36	.47	.305
50,000	.31	.48	.32	.37	.49	.315
51,000	.31	.50	.33	.38	.50	.325
52,000	....	....	....	.39	.52	.34
53,000	....	....	....	.40	.55	.35
54,000	....	....	....	.41	.56	.36
55,000	....	....	....	.42	.59	.37
56,000	....	....	....	.44	.60	.39

Breaking weight of Beam LV = 73,000 lbs.
" " " LVI = 70,000 "

Table U and T show deflections of Canadian Spruce Beams which have been in service.

TABLE U.

Deflections of Beams LVII to LIX.

Loads in lbs.	LVII			LVIII		LIX
	45 ins.	Ends.	45 ins.	45 ins.	Ends.	45 ins. At End.
1,000	.01	.02	.01	.030	.040	.040
1,500	.02	.05	.025	.050	.065	.056
2,000	.035	.07	.05	.060	.100	.070	.09
2,500	.05	.09	.07	.080	.130	.095
3,000	.06	.11	.09	.100	.160	.115
3,500	.075	.14	.10	.120	.190	.130
4,000	.09	.15	.115	.140	.215	.150	.20
4,500	.10	.17	.135	.160	.250	.170
5,000	.115	.20	.15	.175	.270	.190	.25
5,500	.13	.22	.165	.200	.300	.205
6,000	.14	.24	.19	.210	.330	.225	.30
6,500	.16	.26	.20	.240	.360	.248
7,000	.17	.28	.21	.255	.390	.251	.36
7,500	.185	.30	.22	.275	.420	.285
8,000	.20	.33	.235	.300	.450	.305	.41
8,500	.21	.35	.25	.315	.475	.320
9,000	.225	.37	.26	.310	.500	.342
9,500	.235	.39	.275	.350	.535	.362
10,000	.25	.41	.29	.375	.570	.380	.52
10,500	.265	.44	.30	.400	.590	.400
11,000	.275	.46	.315	.410	.620	.415
11,500	.29	.47	.33	.440	.650	.440
12,000	.30	.50	.35	.450	.675	.460
12,500	.32	.52	.36	.475	.705	.480
13,000	.335	.54	.37	.500	.745	.500
13,500	.35	.55	.39	.510	.765	.515
14,000	.36	.57	.40	.540	.800	.540
14,500	.37	.60	.415	.550	.840	.555
15,000	.39	.62	.43	.575	.860	.580
15,500	.40	.65	.45	.600	.900	.620
16,000	.415	.67	.46	.615	.920	.630
16,500	.435	.69	.47	.640	.960	.645
17,000	.45	.72	.49	.655	.990	.665
17,500	.46	.74	.50		1.025	
18,000	.475	.76	.52			
18,500	.50	.78	.54			
19,000	.51	.80	.56		1.120	
19,500	.525	.83	.575			
20,000	.55	.87	.59		1.180	
21,000		.92			1.270	
22,000		.97			1.350	
23,000		1.10			1.430	
24,000		1.50			1.570	
25,000		2.40				
26,000					1.850	
27,000					2.040	

The Breaking weight of Beam LVII = 25,700 lbs.
" " " LVIII = 27,470 "
" " " LIX = 21,700 "

TABLE V.

Deflections of Beams LX to LXI.

Loads in lbs.	LX.			LXI.		
	34 ins.	At End.	34 ins.	46 ins.	At End.	46 ins.
500	.....	.....	.....	.015	.02	.01
1,000	.005	.015	.005	.04	.05	.03
1,500	.005	.015	.015	.06	.09	.05
2,000	.020	.050	.020	.085	.14	.07
2,500	.035	.070	.035	.105	.17	.10
3,000	.045	.080	.045	.135	.20	.12
3,500	.055	.100	.055	.150	.24	.15
4,000	.065	.120	.065	.180	.290	.170
4,500	.070	.140	.070	.20	.320	.190
5,000	.080	.145	.080	.23	.350	.210
5,500	.095	.165	.100	.245	.390	.245
6,000	.105	.185	.105	.265	.430	.260
6,500	.115	.200	.115	.29	.46	.28
7,000	.130	.220	.130	.31	.51	.31
7,500	.140	.240	.145	.34	.54	.335
8,000	.155	.255	.155	.36	.57	.355
8,500	.175	.285	.170	.39	.61	.38
9,000	.180	.300	.185	.41	.65	.40
9,500	.190	.320	.195	.435	.70	.43
10,000	.205	.345	.205	.455	.74	.45
10,500	.220	.365	.220	.49	.76	.485
11,000	.230	.380	.230	.51	.79	.50
11,500	.250	.415	.255	.54	.85	.54
12,000	.....	.440	.....	.....	.92	.....
13,000	.....	.457	.....	.....	.95	.....
14,000	.....	.510	.....	.....	1.03	.....
15,000	.....	.565	.....	.....	1.08	.....
16,000	.....	.610	.....	.....	1.20	.....
17,000	.....	.690	.....	.....	1.32	.....
18,000	.....	.750	.....	.....	1.41	.....
19,000	.....	.870	.....	.....	.....	.....
20,500	.....	.930	.....	.....	.....	.....

Breaking weight of Beam LX = 16,050 lbs.
" " " LXI = 18,400 "

COMPRESSIVE STRENGTH.

The experiments to determine the compressive strength of the various timbers have been chiefly made with columns cut out of the sticks already tested transversely. These columns were, in the first place, carefully examined to see that they had suffered no injury. The following inferences may be drawn:—

(1) The compressive strength of Douglas Fir and of other soft timbers is much less near the heart than at a distance from the heart. Attention may be directed to the case of three equal specimens A, B and C (see photograph page 19), cut out of Beam XIII. The compressive strength of C was found to be 7,706 lbs. per square inch as compared with 6,653 lbs. per square inch, the compressive strength of A. The difference of strength is undoubtedly due to the very much larger proportion of soft to hard fibre, or of summer to spring growth in C, as compared with the proportion in the case of A. The compressive strength of the timber increases with the density of the annular rings.

(2) When knots are present in a timber column, the column will almost invariably fail at a knot or in consequence of the proximity of a knot.

(3) Any imperfection, as, for example, a small hole made by an ordinary cant hook, tends to introduce incipient bending, or crippling.

(4) When the failures of average specimens commence by an initial bending, the compressive strengths of columns of about 10 to 25 diameters in length agree very well with the results obtained by Gordon's formula, the co-efficients of direct compressive strength per square inch being 6000 lbs. for Douglas Fir and 5000 lbs. for White Pine.

Gordon's formula, however, is not at all applicable in the case of specially good or bad specimens. It is often found that a very clear, sound specimen, of even more than 20 diameters in length, will show no signs of bending, but will suddenly fail by crippling under a load as great as that sufficient to crush a shorter specimen.

(5) The greatest care should be observed in avoiding obliqueness of grain in columns, as the *effective* bearing area, and therefore also the strength, are considerably diminished.

(6) If the end bearings are not perfectly flat and parallel, the columns will in all probability fail by bending concave to the longest side.

(7) The *average* strength per square inch, independent of the ratio of length to diameter, is:

5974	lbs. for New Douglas Fir	
6265	" for Old "	"
4067	" for New Red Pine	
3843	" for New White Pine	
2772	" for Old "	"
3617	" for New Spruce (B.C.)	
5136	" Old Spruce	

It should be pointed out that none of the old Douglas Fir columns exceeded 4.4 diameters in length, while the great majority of the new Douglas Fir columns were from 4 to 25 diameters in length. This explains the reason of the greater average compressive strength of the old Douglas Fir. A similar remark applies to the New and Old Spruce.

Table giving in detail the results of the experiments on the different specimens:—

RESULTS OF COMPRESSION TESTS ON NEW DOUGLAS FIR.

Dimensions in ins. Lengths.			Breaking Load in lbs. per sq in.	Weight in lbs. per cub. ft.	Remarks.
3.07 ×	3.08 ×	3.11	6367		Failed by bulging.
3.06 ×	3.08 ×	3.10	5760		Failed by folding.
2.63 ×	3.63 ×	5.81	4923	30.3	Specimen 3" or 4" from heart; grain straight; one small knot on high edge. Failed by crippling at knot on high edge.
3.65 ×	3.65 ×	6.12	3678	29.8	Heart piece; grain straight but seasoned; annular rings very wide; two knots, one on high edge. Failed at this latter by crippling.
2.19 ×	3.74 ×	5.40	4761	38.4	Straight grained; one large knot from side to side; specimen 3" or 4" away from centre. Failed at knot.
4.10 ×	4.30 ×	8.05	5218	32.9	Large knot on one end; many small knots all through piece; also heavy season cracks. Failed by bursting along season cracks and through knots.
2.15 ×	2.25 ×	9.2	5809	38.8	All clear. Failed by crippling.
2.17 ×	2.25 ×	9.14	7313	35.1	Sound, clear and straight grained; small deficiency on one side at end. Failed by crippling.
2.12 ×	2.16 ×	9.15	7294	38.7	Straight grained; clear on three sides; 4th side old, with bad defect 4 ins. from one end. Bulged and failed at defect.
2.22 ×	2.22 ×	9.07	8177	37.5	Straight grained and clear; one bad season crack. Failed by crippling.
2.13 ×	2.20 ×	9.15	6850	36.5	Straight grained; small knot near one corner 3 ins. from end. Failed at this knot.

RESULTS OF COMPRESSION TESTS ON NEW DOUGLAS FIR.—*Continued*.

3.32 ×	3.32 ×	9.62	3810	29.5	Heart piece; straight grained; two heavy season cracks; three or four pin knots. Failed by bulging on season cracks; and crippling through two pin knots on same side.
3.33 ×	3.34 ×	10.58	4388	33.0	Clear; straight grained. Failed on high side. Specimen 3" or 4" from heart.
3.45 ×	3.50 ×	10.60	7000	32.6	Clear and straight grain; somewhat shaken; crippled 6 ins. from end.
2.74 ×	4.27 ×	11.25	6837	35.3	Clear and straight grained; some season cracks; failed by crippling directly across about 1½ ins. from one end.
2.85 ×	4.25 ×	11 27	5615	30.0	Clear and straight grained, but season cracks along annular rings, and one heavy season crack along medullary rays. Failed first by bursting apart of piece at a season crack, then by crippling of the remainder.
3.94 ×	3.95 ×	11.97	7069	33.8	Clear straight grain; season crack on one side. Failed by crippling at middle on the highest edge.
2.72 ×	2.92 ×	11.85	8942	40.0	Clear and straight grain; shaken over 8 ins. crippled 4 ins. from end.
3.46 ×	3.48 ×	12.04	5481	30.4	Two sets of knots, one at one end, the other at centre. Failed at both by crippling, at same time.
4.05 ×	4.10 ×	12.01	5542	35.1	Knots (heavy) on one end; also several near other end; grain curved at various places due to knots. Grain bent at knot at end.
2.85 ×	3.75 ×	12.5	6155	38.3	All clear. Failed by crippling.
2.92 ×	3.79 ×	12.5	5966	39.3	All clear. Failed by crippling.
2.9 ×	4.37 ×	12.0	6265	35.5	One old side; grain straight and parallel; one side inclined 1-in. in 12 ins.; on other side, two season cracks. Failed by crippling.
2.79 ×	3.43 ×	12.0	5363	35.7	One old side; grain straight and nearly parallel; no seasoning cracks. Failed by crippling.
2.92 ×	4.42 ×	12.0	5262	34.2	One old side, grain straight and parallel; one season crack. Failed by crippling.
2.87 ×	3.39 ×	12.0	6784	35.1	Two old sides; grain nearly parallel; no season cracks. Failed by crippling.
2.93 ×	3 42 ×	12.03	5520	33.9	Clear and straight grained; one old side with deep seasoning cracks; a slight crack through centre of piece. Crippled 4 ins. from end, and bulged along season crack.
2.80 ×	4.40 ×	12.0	5060	36.4	Straight grained; one old side with many season cracks. Failed by splitting down season cracks and afterwards crippling.

RESULTS OF COMPRESSION TESTS ON NEW DOUGLAS FIR. —Continued.

2.78 × 4.38 × 12.0	6500	35.5	Straight grained and clear; one old side with season crack nearly across piece. Crippled 3 ins. from one end.
2.82 × 3.48 × 12.02	6010	35.9	Grain straight; two old sides; piece sound, no flaws. Crippled near one end.
3.3 × 3.98 × 12.0	5560	34.2	Grain straight and clear, except small pin knot on a corner 4 ins. from end; had two bad season cracks the whole length. Crippled 4 ins. from end induced by season cracks; also bulged out.
3.38 × 3.43 × 13.53	6816	34.7	Clear; grain bent out of straight at one end, due to proximity of knot, also somewhat shaken. Failed by bursting along fibres out of parallel.
2.20 × 2.24 × 13.78	5638	34.3	Grain out of parallel for 1 in. in length; knot on one corner of end. Burst along shaken fibres out of parallel.
3.38 × 3.45 × 13.90	6861	33.8	Straight grained, except one-half of a knot on one end. Failed by crippling near knot at end.
4.03 in diar. × 48.01	5856	31.3	Grain parallel, no knots; two small cracks and a small split; annular rings nearly straight. Failed by bending concave to a high corner.
2.84 × 4.23 × 13.12	5828	31.5	Straight grained, small pin knot 3 ins. from one end; season cracks from end to end through middle, passing through knot. Failure by opening of season cracks, and crippling through knot.
4.10 × 4.45 × 14.47	7188	39.1	Clear; grain out of parallel. Failed by crippling and shearg in of unsupported fibres.
2.70 × 2.90 × 15.96	8365	39.5	Clear, straight grain shaken over a length of 11-ins. Crippled 5 ins. from end.
2.16 × 2.20 × 16.29	6442	36.0	Clear, not straight grain; somewhat shaken; sheared along shake in grain which being cut off parallel had no bottom support.
4.08 in diar. × 24.12	6595	31.8	Clear and straight grained. Failed by crippling 10 ins. from end.
2.70 × 4.20 × 16.45	6349	30.8	Straight grained; season cracks on one side; several small pin knots. Failed by crippling 2 ins. from one end through one of the pin knots.
2.38 × 3.56 × 16.74	7143	33.0	Straight grain; some small pin knots. Crippled through the largest one at centre.
1.73 × 5.98 × 17.73	4209	38.7	Grain parallel knot on edge 4 ins. from end; also bad season crack and small deficiency in one corner for 6 ins. from one end. Burst at knot and split along season crack.

RESULTS OF COMPRESSION TESTS ON NEW DOUGLAS FIR.—*Continued*.

17 ×	2.25 ×	17.12	7700	35.6	Clear, straightgrained Failed by bending and crippling 3 ins. from end.
3.11 ×	4.00 ×	17.49	4702	33.2	Two heavy knots at centre, one running from side to side through centre; grain crooked and not parallel. Failed by grain shearing and bursting through knot at centre.
3.12 ×	4.03 ×	17.70	4217	34.2	One heavy knot at centre running from corner to corner, other smaller knots; grain crooked and out of parallel. Crippled at knot at centre.
1.75 ×	5.82 ×	17.79	5135	37.8	Grain straight and sound; season cracks in centre. Failed by crippling at both ends and also by bending, which probably first caused failure.
3.95 ×	5.81 ×	17.80	6432	39.1	Grain clear and straight, but not parallel; slight season cracks. Failed by cripple across 4 ins. from one end.
3.95 ×	5.92 ×	17.82	5359	38.0	Grain clear and straight; some season cracks. Crippled 6 ins. from end.
4.97 ×	4.95 ×	17.83	4504	37.9	Grain straight and parallel; bad knot 7 ins. from end passing through piece. Failed by bursting at knot and along grain.
1.71 ×	5.95 ×	17.84	5464	36.0	Grain parallel and clear; bad season crack through heart. Failed by bending at centre. Crippled on concave side.
1.79 ×	6.00 ×	17.85	6034	36.3	Grain straight and clear; bad season cracks; also chip out on a corner 4 ins. from one end. Failed at round end by crippling and by opening of season crack.
3.95 ×	5.95 ×	17.89	6225	38.9	Clear and straight grained; slight season checks. Crippled 3 ins. from one end.
4.08 ×	4.45 ×	19.68	6437	36.7	Clear, but badly out of parallel. Failed by bursting along fibres out of parallel.
3.02 ×	4.04 ×	19.97	3240	30.8	Two heavy knots at centre, one also at one end, several other smaller ones. Failed by bursting down centre through knots.
3.85 ×	3.91 ×	24.05	5382	35.2	Grain straight; two knots on adjacent sides, one at 8 ins. from each end; season cracks running diagonally at one end. Failed by crippling at large knot.
4.35 ×	4.85 ×	20.75	3630	28.0	Failed by shearing and crippling; grain clear, but not quite parallel.
2.20 ×	2.24 ×	21.05	7424	35.0	Clear, and straight grained; tested before as pillar. Failed by bending 4 ins. from end.
2.92 ×	3.30 ×	24.27	4606	34.6	Straight grain; knot 6 ins. from end passing through a corner. Crippled at knot.

RESULTS OF COMPRESSION TESTS ON NEW DOUGLAS FIR.—*Continued.*

2.60 ×	3.23 ×	25.4	4416	34.7	Straight grain; large knot 4 ins. from end on an edge. Failed by crippling at knot.
2.27 ×	2.28 ×	23.46	4363	36.91	Straight grained; clear except part of knot on one end. Failed by crippling at knot.
4.20 ×	4.36 ×	27.88	2622	32.4	Heart; grain 2½ ins. out of straight; heavy season cracks; two large knots. Failed by bulging along season crack and at knots 14 ins. from end.
4.05 ×	4.20 ×	24.70	5026	33.9	Tested before as pillar, failed then at 67,200 lbs. This portion had straight grain; two knots close together 8 ins. from one end going through piece. Failed by crippling at these knots.
2.61 ×	2.65 ×	24.42	6237	36.0	Straight grain; season crack across end running half the length of the piece; knot 3 ins. from other end ½ in. in diameter. Crippled at the knot.
2.65 ×	2.66 ×	26.24	6865	36.4	Straight grained and clear; season crack running down about 8 ins. Crippled clean across at foot of season crack, apparently not induced by seasoning.
2.00 ×	2.01 ×	27.40	6841	34.5	Clear and straight grain; heavy season crack. Burst from end to end on season crack.
2.88 ×	2.95 ×	23.91	8106	38.8	Clear, straight grained. Crippled 8 ins. from one end.
2.87 ×	2.93 ×	25.00	6600	35.5	Clear, nearly straight grained; slight season crack. Failed by a bulging on season crack and afterwards crippled on reduced section at centre.
2.88 ×	2.90 ×	24.40	7856	36.4	Clear, straight grained. Failed by direct cripp'g.
2.87 ×	2.90 ×	24.55	8065	38.0	Clear and straight grained. Failed by direct crippling 8 ins. from end.
2.90 ×	2.95 ×	25.70	8023	36.3	Clear and straight grained. Failed by direct crippling 15 ins. from end.
2.78 ×	2.87 ×	25.95	9700	40.9	Deficiency near centre, about ½ in. by 1 in. (resin); fibre crooked through vicinity of knot; otherwise clear and straight grained. Failed at crooked fibres at deficiency.
2.89 ×	2.90 ×	26.69	8269	33.4	Clear and straight grained; failed by compression of fibres on a corner.
2.82 ×	2.97 ×	25.15	9104	40.2	Very heavy summer rings; clear; fibres bent 12 ins. from one end at one side due to vicinity of a knot. Failed at crooked fibres.
4.77 ×	5.82 ×	26.15	7709	36.5	Did not fail.
4.77 ×	4.68 ×	22.32	8411		Same as preceding with piece cut off; clear and straight grain.

RESULTS OF COMPRESSION TESTS ON NEW DOUGLAS FIR.—Continued.

Dim 1	Dim 2	Dim 3	Load	Strength	Notes
4.70 ×	5.85 ×	25.78	6653	29.2	Straight grained; one knot from side to side at centre. Failed by crippling and bulging at knot.
2.27 ×	2.27 ×	31.0	3823	37.2	Grain not straight; one pin knot; also knot on one edge 12 ins. from end. Failed by bending at knot on high corner.
3.38 ×	4.33 ×	32.20	6425	41.3	Clear, straight grained. Crippled 1 ft. from end.
3.39 ×	4.42 ×	30.90	5935	37.8	Clear, straight grained; external fibre burst; then crippled near centre.
3.38 ×	4.42 ×	32.32	6111	43.3	Clear, straight grained; burst, then crippled at centre.
3.37 ×	4.38 ×	32.5	5420	38.9	Clear, straight grained; season crack on one side; small season crack across end. Crippled near end.
3.35 ×	4.36 ×	31.55	6486	43.1	Clear and straight grained. Crippled near end.
3.41 ×	4.45 ×	32.4	5880	37.6	Clear and straight grained. Crippled near end.
3.27 ×	3.42 ×	31.75	5760	33.5	Straight grained; knot ½-in diar., from side to side. Failed by crippling at this knot 8 ins. from one end.
2.65 ×	2.86 ×	30.65	5047	36.3	Clear, straight grained. Failed by crippling 8 ins. from one end.
2.67 ×	2.88 ×	31.83	7607	35.3	Clear straight grained. Failed by crippling and bending at same instant at centre.
3.28 ×	3.45 ×	33.81	6940	35.7	Clear, and straight grained. Failed by bending 10 ins. from one end.
2.75 ×	2.82 ×	30.47	5480	33.0	Nearly straight grained; various small knots, one larger knot ⅜ in. diar. 3 ins. from one end. Failed by crippling at this knot; also somewhat sea-oned at heart.
2.90 ×	2.90 ×	29.35	6183	32.7	Straight grained; various small knots, one larger knot ⅜ in. diar. 9 ins. from end. Failed by crippling at this knot.
2.75 ×	2.88 ×	31.50	5871	36.4	Straight grained; knot ¾ in. diar. 12 ins. from end. Crippled at the knot.
2.17 ×	2.18 ×	30.00	6174	35.0	Straight grained, clear but for one knot 10 ins. from end ½ in. in diar., Crippled at this knot.
2.73 ×	2.85 ×	28.74	8124	34.8	Clear and straight grained. Failed by a thin layer bursting out, and then a clean cripple 8 ins. from same end.
4.69 ×	5.84 ×	28.10	6677	31.1	Clear and straight grained; crippled 8 ins. from end.
4.17 ×	5.00 ×	33.70	4839	32.3	Straight grained, but heavy knot near end and very heavy knot near centre. Crippled at latter knot.
4.30 ×	5.01 ×	32.72	5566	36.7	Straight grained, but heavy knot on side near centre; also heavy knot 8 ins. from end one side. Failed at the latter knot.

RESULTS OF COMPRESSION TESTS ON NEW DOUGLAS FIR.—*Continued.*

Dimensions	Load	Strength	Remarks
3.95 × 4.33 × 32.28	4479	30.1	A great many knots on each end and at various other points. Failed at a large knot 12 ins. from an end. Also heavy season cracks.
3.98 × 4.10 × 28.65	5735	34.3	One old side badly seasoned and injured by useage; also knots near each end; also a small pin knot near centre at which piece failed by crippling and bursting of fibres.
3.93 × 4.30 × 31.95	5124	32.6	Heavy knots near centre. Crippled at knots.
4.11 × 4.92 × 31.85	7309	35.1	Clear and straight grained, except slight wave 1 ft. from end due to vicinity of knot. Failed at this point by direct crippling.
4.22 × 4.92 × 30.84	7167	39.2	Clear and straight grained. Crippled 8 ins. from end.
2.33 × 2.84 × 28.00	6496	31.7	Clear and straight grained. Failed by bending 10 ins. from end.
2.27 × 2.27 × 33.75	5708	36.0	Clear, straight grained. Failed by bending; short specimen failed at 30,000 lbs.
3.96 × 4.18 × 35.25	5015	36.6	Several knots; crippled at one running from corner to corner 12 ins. from one end.
4.20 × 4.50 × 38.00	5905	35.6	Grain out of parallel; clear. Failed by bursting and shearing along season cracks.
3.33 × 3.40 × 33.55	7615	33.6	Clear, straight grain. Crippled near one end.
3.30 × 3.38 × 33.54	7444	35.6	Clear and straight grained. Failed by crippling 6 ins. from end.
3.35 × 3.40 × 33.50	5338	35.4	Large knot passing through centre side to side; piece split end to end through this knot.
3.30 × 3.40 × 33.55	5909	35.6	Knot near centre, also two small pin knots near end. Crippled through pin knots.
3.30 × 4.00 × 33.50	5416	35.2	Large knot near centre passing from side to side. Split from end to end through knot.
3.30 × 4.00 × 33.50	5023	32.8	Large mass of knots near middle. Crippled at these.
4.25 × 5.75 × 35	5729		
4.25 × 5.87 × 41.75	4000		
4 × 4 × 48	4469	32.75	Grain parallel; knot at centre at corner; other knots near end; centre of tree 12 ins. away. Bent at centre at knots concave to a high corner.
2.86 × 4.06 × 40.02	6330	38.1	Straight grain; small knot 14 ins. from end. Failed by bending in middle.
4.10 × 4.24 × 41.83	3866	36.3	Straight grain; three knots. Crippled at knot 12 ins. from end; no bending.
4.25 × 4.25 × 54.95	3389	34.6	Straight grain; many knots. Burst in two opposite directions at knots 11 ins. from one end and 12 ins. from other end.
1.99 × 2.64 × 52.62	5105	34.3	Straight grain; clear; bent at centre.

RESULTS OF COMPRESSION TESTS ON NEW DOUGLAS FIR.—*Continued.*

Dimension in ins.		Lengths.	Breaking Load in lbs. per sq in.	Weight in lbs. per cub. ft.	Remarks.
4.26 ×	4.33 ×	60.0	3980	35.5	Straight grain; failed by crippling at knot passing through corner 13 ins. from end and 1-16 in. out of square; no appreciable effect.
4.09 ×	4.34 ×	59.0	3211	34.4	Straight grain; three or four knots; season crack on one side. Crippled at knot 20 ins. from end and season crack opening.
4.18 ×	4.22 ×	59.75	3190	35.4	Four knots, two each 18 ins from each ends, several other small knots; grain not straight; large season crack. Failed by shearing and bursting open at season crack across annular rings.
2.46 ×	2.51 ×	60.5	4619	34.5	Straight grain; several knots. Failed by crippling at knot 12 ins. from end.

RESULTS OF COMPRESSION TESTS ON OLD DOUGLAS FIR.

Dimension in ins.		Lengths.	Breaking Load in lbs. per sq in.	Weight in lbs. per cub. ft.	Remarks.
2.21 ×	2.23 ×	9.15	8644	35.9	Grain straight and clear; one old side with season crack. Bulged along season crack, and crippled.
3.45 ×	2.78 ×	9.65	6465	32.5	All fresh sides; straight and parallel grain; one edge strained from bolt. Crippled all over.
3.41 ×	2.78 ×	9.65	7247	35.4	One old side; grain straight and parallel. Crippled near one end.
3.41 ×	2.80 ×	9.70	5696	33.2	All fresh sides; grain straight and parallel; one edge strained from bolt; 1 in. season crack. Crippled one-fourth the way down, slightly helped by season crack.
3.38 ×	2.78 ×	9.65	6979	34.5	One old side; grain straight and parallel. Crippled at one end, slightly aided by season crack.
2.76 ×	3.76 ×	9.64	7235	35.6	One old side; iron stain at one end; season crack; grain straight and parallel. Crippled at 3 ins. from end.
2.83 ×	3.81 ×	9.75	6577	32.9	One old side; grain straight and parallel. Crippled near centre.
4.15 ×	4.64 ×	11.32	6660	35.70	Knot 5 ins. from end; next face, knots 1½ ins. and 4 ins. from same end; small pin knot and season crack on third side. Crippled through knots.
4.35 ×	4.67 ×	11.95	7900	47.25	Clear and straight; very full of resin; some season cracks; crippled at one end.
3.40 ×	3.47 ×	12.00	5085	31.7	Grain straight, but slightly curly; three fresh sides; old side crushed by tie; slightly rotten under tie; crippled at small defect near one end.
3.45 ×	3.45 ×	12.00	5218	30.88	Grain parallel; crushed and rotten for a depth of ⅜ in. under tie; two adjacent sides new. Crippled at rotten part near one end.

RESULTS OF COMPRESSION TESTS ON OLD DOUGLAS FIR.—Continued.

Dimensions	Load	Strength	Remarks
3.45 × 3.47 × 12.0	3838	35.0	Grain parallel, but crooked; knot near corner 4¼ ins. from end, 1½ ins. diar., knot extended into piece. Crippled through knot.
3.45 × 3.47 × 12.0	4928	38.7	Grain parallel; three fresh sides; 1¾ ins. knot passing through corner 5 ins. from end. Crippled near one end and split along grain adjacent to knot.
3.45 × 3.45 × 12.0	5461	33.3	Grain parallel; two adjacent fresh sides; season crack on one old side. Crippled near one end and split slightly along season crack.
2.90 × 2.92 × 12.0	5314	34.0	Grain parallel; three fresh sides; small season crack. Crippled near one end.
3.41 × 3.48 × 12.0	5308	34.9	Grain parallel; three fresh sides; knot hole on one corner 3½ ins. long, 0".8 in. deep; also season cracks. Failed by opening of season cracks.
3.42 × 3.47 × 12.0	4011	30.0	Grain parallel; three fresh sides; old side slightly damaged; also cant. hook holes. Crippled near centre at cant. hook holes.
3.42 × 3.45 × 12.0	4814	32.0	Grain parallel; two fresh sides; slightly rotten at one end on old side. Crippled at the rotten point.
3.45 × 3.46 × 12.0	5053	30.5	Straight grain; all fresh sides; shows signs of failure; crack at end. Crippled near one end.
2.88 × 2.87 × 12.0	6199	33.2	Grain sound and parallel; three fresh sides. Crippled near one end.
3.44 × 3.46 × 12.0	5703	33.6	Grain parallel; two adjacent fresh sides; season cracks; small cant. hook hole 2 ins. from end close to corner; slightly rotten. Crippled at cant. hook mark.
3.46 × 3.46 × 12.0	5093	33.8	Grain parallel; three fresh sides; small season crack on one side. Crippled at one end; season crack opened.
2.82 × 3.40 × 12.05	6611	32.7	Parallel grain; four fresh sides. Crippled near one end.
2.77 × 3.36 × 12.0	7519	35.3	Parallel grain; one old side; saw cut and season crack. Crippled near one end.
2.80 × 3.40 × 12.03	6813	32.5	All fresh sides; grain straight and parallel; 1 in. season crack. Split along season crack.
2.79 × 3.35 × 12.03	6845	34.6	One old side; season cracks; grain straight and parallel. Split along season crack.
2.79 × 3.91 × 12.03	7149	34.6	One old side; grain straight and parallel. Crippled at one end.
2.78 × 3.73 × 12.04	7348	35.5	One old side; grain straight and parallel; season cracks 1 in. deep. Crippled at one end.
2.77 × 3.86 × 12.05	7390	33.5	One old side; grain straight and parallel. Crippled near centre at a small defect.

RESULTS OF COMPRESSION TESTS ON OLD DOUGLAS FIR.—*Continued.*

Dimensions	Load	Strength	Remarks
2.80 × 3.80 × 12.06	7481	31.1	One old side; grain straight and parallel. Crippled at end.
2.78 × 3.88 × 12.0	7090	34.2	One old side; grain straight and parallel. Crippled near one end.
2.79 × 3.06 × 12.0	7317	33.4	One old side; grain straight and parallel. Crippled at 3 ins. from end.
3.27 × 3.95 × 12.0	6540	33.45	Grain straight and clear, except small pin knot hole 3 ins. from end; piece shivered by season cracks. Failed by piece splitting off. It then crippled at knot 3 ins. from one end.
3.28 × 3.96 × 12.	5510	32.9	Grain straight; small pin knot on a corner near centre; very heavy season crack on old side. Burst along season crack; also crippled 4 ins. from one end.
3.32 × 4.04 × 12.0	4825	28.85	Grain straight; pin knot on corner near centre; heart decayed; also one season crack. Crippled at pin knot.
3.31 × 4.02 × 12.04	5675	32.85	Grain straight; small pin knot 1½ ins. from end; two bad season cracks. Crippled square across near each end.
3.33 × 4.0 × 12.0	4165	28.95	Grain not quite straight; knot at corner 2 ins. from end; deficiency of heart all along one edge. Crippled at knot.
3.30 × 4.0 × 12.0	6300	33.55	Straight grain; knot on corner 1½ ins. from end; large deficiency on opposite corner at other end; another deficiency and nail gouge at centre of same edge; also one season crack. Crippled at knots.
3.28 × 4.02 × 12.03	5540	32.70	Straight grain; knot on corner 1½ ins. from end; also season cracks. Crippled 4 ins. from end.
4.18 × 4.63 × 12.22	5200	35.3	Knots 3 ins. and 6 ins. from end on same side; also small knot on next face 1 in. from same end; also part of large knot on other end. Failed longitudinally through two knots; upper end was not horizontal, not more than 5·6 ths. of the area bearing.
4.35 × 4.65 × 14.15	6735	36.95	Two knots 2 ins. and 6 ins. from end on same side; also knot on next face 3 ins. from same end and two knots on other end; on third and fourth faces, knots 1½ ins. and 4 ins. from first end. Crippled at knot 3 ins. from end.
4.25 × 4.65 × 14.80	7085	36.6	Two knots passing through from face to next face; one 3 ins. from end; the other 7 ins. from same end; deficiency 1 in. × 1¼ in. on opposite edge. Crippled through knot 7 ins. from end.

RESULTS OF COMPRESSION TESTS ON OLD DOUGLAS FIR.—*Continued.*

4.39 × 4.70 × 14.78	6500	45.70	Full of resin; part of large knot on one end; season crack on one face; shaken on a corner. Crippled in solid wood (in resin part) 4 ins. from end.
4.14 × 4.65 × 14.80	6730	41.0	Patch of resin through centre; knot on one corner 6 ins. from end; slight season cracks; slight deficiency on one corner. Crippled through knot.
4.25 × 4.66 × 14.78	6020	37.4	One medium knot 1 in. from end; also many small knots on same face; on next face, knots at 6 ins. and 1 in. from same end. Failed through knots at the centre.
4.16 × 4.60 × 14.50	7410	35.7	Part of large knot on one end; one side covered with small knots; otherwise sound specimen. Failed at large knot at end.
4.28 × 4.70 × 14.78	7490	36.2	Grain parallel; one medium knot 5 ins. from end; also two small knots 1 in. from same end and on same side; also heart shake. Failed at centre by crippling through small knot.
4.17 × 4.70 × 14.78	6400	34.0	Grain parallel; mass of knots at one end; also badly seasoned in resinous portion. Crippled at knotty end.
4.35 × 4.74 × 14.80	6310	47.0	Grain parallel; large knot near one end; bad season cracks in resinous portion. Crippled at large knot.
4.27 × 4.67 × 14.80	7310	37.2	Grain straight and sound, but one large knot on end; also one knot on an edge 3 ins. from end; one knot 5 ins. from other end on same edge; slight season cracks. Failed at the two last knots.
4.14 × 4.57 × 14.75	6960	35.45	Knots in each end; otherwise clear; two old sides badly shaken. Crippled and burst at knot at one end.
4.32 × 4.70 × 14.80	5970	38.05	Groups of small knots about 3 ins. from each end; also full of resin. Crippled at each end through knots.
4.14 × 4.60 × 14.80	6580	35.05	Groups of small knots about 4 ins. from each end; also bad season cracks. Crippled through one group of knots.
4.06 × 4.65 × 14.85	6500	43.70	Large knot at one end; two knots 5 ins. from other end; full of resin; dense and heavy; one season crack. Crippled through both knots 5 ins. from end.

RESULTS OF COMPRESSIVE TESTS ON RED PINE.

Dimensions in inches.	Lengths in inches	Compressive Strength in lbs. per sq. inch.	Weight in lbs. per cub. ft.	Remarks.
4.96 in diar. × 5.9		2497		Failed at knots 26 ins. from end; also at another ring of knots 3 ins. from same end; nineteen knots in length.
4.97 in diar. × 5.8		2742		
2.98 " × 5.86		2722		
3.00 " × 5.9		2631		
2.95 in diar. × 5.65		6870		One knot near one end. Failed by crippling above knot.
2.88 in diar. × 5.69		7057		Clear. Crippled 6 ins from one end.
4.81 in diar. × 13.75		5092		Clear grain. Failed by spreading at bottom.
3.88 in diar. × 13.5		7602	39.9	Nearly straight grain; knot 6 ins. from end passing nearly through centre. Failed at the knot by crippling.
3.80 diar. × 13.31		6438	35.8	Straight grained; knot on one end. Failed by crippling at knot about ½ in. from end all around
4.02 in diar. × 18.75		4657		Clear wood; straight grained; spread at end, due to curvature of fibre in locality of a knot.
3.90 " × 18.20		7222	35.7	Clear and straight grained. Failed 6 ins. from end by folding.
3.66 " × 22.61		8516	43.2	Grain parallel; one knot 10 ins. from end. Failed through knot by crippling.
4.01 in diar. × 22.73		5637	28.7	Four knots at 8 ins. from one end. Failed by crippling at knots.
4.3 in diar. × 22.8		5983	26.7	
3.93 in diar. × 29.2		7911	38.1	Grain parallel; two knots, one large knot 10 ins. from one end. Failed by crippling at this knot.
6.93 " × 36.12		2698		Failed by crushing at knot, 4 ins. from end. Fourteen knots in length.
7.02 " × 36.12		2087		Failed at knot 8½ ins. from end; ten knots in length.
7.01 " × 36.12		2024		Failed at ring of knots 7 ins. from end; fifteen knots in length.
3.97 " × 3.10		3287		Crushed and failed at knot; straight grain; fairly free from knots.
4.10 " × 3.10		2825		Failed by crushing and bending. Straight grain; crack down length.
4.04 " × 3.10		3482		
4.03 " × 3.10		4247		
3.98 " × 3.10		3223		
3.96 " × 3.10		4001		
4.75 × 4.75 × 60.		3104		
3.97 in diar. × 69.		2585	.985	Not well seasoned. Failed by crushing and bending at a large knot 31 ins. from end; also at 1 in. from end and 4½ ins. from other end; straight grained; six knots in whole length.
4.08 " × 69.		2593		Failed at ring of knots four in number by crushing and bending at 24 ins. from end; also at 2 ins. from same end; fourteen knots in whole length.

RESULTS OF COMPRESSION TESTS ON RED PINE.—*Continued.*

Dimensions			Compressive Strength lbs/sq.in	Remarks
4.02 in diam.	×	69	3152	Failed by crushing; straight grained; failed at two small knots 27 ins. from end and also at 16 ins. from same end; large knots 39 ins. from same end; ten knots in length.
3.91	" ×	69	3280	Failed by crushing 16 ins. from one end at a knot. Twelve knots in whole length.
4.03	" ×	69	3158	Failed chiefly by crushing 12 ins. from one end; four knots in length.
3.96	" ×	69	3734	Failed at knot 24 ins. from end; six knots in length; also crippled 1 inch from same end.
4.94	" ×	66.25	2386	Failed at knots 26 ins. from end; also at another ring of knots 3 ins. from same end; nineteen knots in length.
4.92	" ×	66.25	2513	Failed at ring of knots 36 ins. from end; sixteen knots in length.
2.96	" ×	66	1977	Failed by crushing and bending at large knot 28 ins. from end, eight knots in length.
3.06	" ×	66.25	2433	Failed by crushing at knots 5 ins. from end Four knots in whole length.

RESULTS OF COMPRESSIVE TESTS ON NEW WHITE PINE.

Dimensions in inches			Lengths in inches	Compre've Strength in lbs. per sq. inch	Weight in lbs. per cub. ft.	Remarks
4.187 ×	2.14	×	2.31	3810		
4.687 ×	2.312	×	2.44	2955		
4.812 ×	2.312	×	2.44	4248		
3.0 ×	2.94	×	2.98	5352	24.4	
4.75 ×	4.75	×	3.	3821		
4.8 ×	4.8	×	4.6	3515		
4.75 ×	4.75	×	4.6	4387		
4.75 ×	4.80	×	4.53	3280		
4.75 ×	4.44	×	4.50	3449		
4.75 ×	4.78	×	4.36	4361		
4.75 ×	4.75	×	4.37	4433		
4.75 ×	4.75	×	4.40	4363		
4.75 ×	4.70	×	4.50	3449		
4.75 ×	4.80	×	4.53	3193		
4.75 ×	4.75	×	4.37	3972		
4.75 ×	4.75	×	5.	3548		
4.75 ×	4.75	×	10.375	2926	30.3	Grain clear but not straight. Cracked down one side.
3.01 in diam.		×	11.35	4382	26.7	Clear and straight Failed by folding near one end.
4.75	"	×	11.125	3500	21.60	Clear grained, but not straight. Failed by folding over at top.
4.75	"	×	11.875	5527	27.50	Clear specimen; deep season cracks across annular rings. Failed by crippling.
4.812	"	×	12.25	3990	23.80	Two large knots. Failed between them.
3.00	"	×	12.80	3762	29.4	Two heavy knots 2 ins. from end. Failed by crippling at the knots
4.75 ×	4.75	×	12.156	5383	26.5	Clear specimen. Crippled without bulging or cracking.
2.98 ×	2.98	×	12.0	5574	29.4	Clear and straight grained. Failed by crippling.

RESULTS OF COMPRESSIVE TESTS ON NEW WHITE PINE.—Continued.

Dimensions	Load	Value	Remarks
4.74 in. diar. × 13.12	2774		Ring of four knots 6 ins. from one end. Failed by crippling at knots.
4.71 in. diar. × 14.562	3400	20.6	One knot and also signs of decay. Failed by crippling at the knot.
2.625 × 3.502 × 14.125	6400		Clear.
4.72 × 4.72 × 14.875	5004	26.3	Clear. Crippled without cracking or bulging.
4.75 in. diar. × 14.75	4408		
4.71 in. diar. × 15.5	3360	21.1	One large knot; decayed near heart. Failed at knot.
4.703 " × 15.35	3861	26.60	One knot at bottom of specimen. Failed at this knot by crippling.
2.94 " × 15.30	4272	26.5	Clear and straight, but deep injury from pike pole. Failed at injured part.
4.75 in. diar. × 16.	4463		
3.87 " × 16.25	2973	29.9	Straight grained. Failed at one end at a large knot.
4.75 " × 17.35	4232	26.40	Two large knots. Failed between them.
4.71 " × 17.938	4847	27.1	Clear and straight grained. Failed at end
4.40 × 4.40 × 17.0	3856	30.6	Three large knots in a ring around specimen. Failed at knots.
2.97 × 3.85 × 20.54	6036	30.1	Clear and straight grained; one-third sapwood. Failed by crippling at 7 ins. from one end.
3.85 × 3.83 × 21.65	3933	26.1	Failed previously as pillar under 49,200 lbs. Crippled now at a large knot 8 ins. from end.
3.8 × 3.8 × 22.35	3908	26.7	Two large knots Crippled at one, 2 ins. from an end.
3.83 × 3.83 × 23.82	3615	25.9	Failed by crippling at two knots near centre.
2.97 × 2.99 × 23.60	5462	24.9	Clear and straight grained; failed previously as pillar under 42,000 lbs. Crippled now near centre.
3.02 diar. × 25.79	5023	24.5	Clear and straight grained. Failed by crippling 8 ins. from one end.
3.40 × 3.80 × 25.4	3610	25.0	Straight grained; bad season cracks; full of knots, failed by crippling through two of them 8 ins. from end.
2.98 × 2.99 × 24.25	4607	23.9	Straight grained; pin knot 10 ins. from one end. Failed by crippling and bending at pin knot.
2.95 × 3.25 × 26.70	3508	24.1	Straight grained, but full of knots. Crippled at one near corner in middle.
4.75 × 4.75 × 21.0	3103		Clear; grain 2 ins. out of parallel; season cracks along grain. At upper corner grain ran out. Failed by sliding along seasoning, due to non support of fibres running from corner.
2.99 × 2.99 × 24.08	4474	26.7	
3.05 in. diar. × 24.1	5240	25.8	Clear and straight grained. Failed by crippling and bending at same instant at middle.
3.46 × 4.33 × 27.00	3488	20.4	Failed previously as pillar under 33,300 lbs. Failed now at knot 8 ins. from end on a side.
2.92 in. diar. × 36.53	5269	29.8	Clear and straight grained; one-third sapwood. Fail by crippling on sapwood side and then bending afterwards 12 ins. from end.

RESULTS OF COMPRESSIVE TESTS ON NEW WHITE PINE.—*Continued.*

Dimensions in inches. Lengths.			Comp've Strength in lbs. per sq. inch.	Weight per cub. ft. in lbs.	Remarks.
3.05 in diar.		× 48.0	4377	25.9	Clear grain, 1½ in. out of straight; high at one side. Failed by bending 20 ins. from one end on high side.
3. ×	3.	× 48 0	4666	25.0	Ten knots; long season crack ran three fourths the way down, 1½ ins. deep and ½ in. from edge; a bruise 3 ins. from end on same side; on opposite side, crack 3 ins. long, 1 in. deep; grain and rings both parallel. Failed by bending toward a high corner and then crippling.
4.75 in diar.		× 60	2652		
4.75 in diar.		× 60	1862		
4.75 ×	4.75	× 60	2749		
4.75 ×	4.75	× 60	1862		
4.75 ×	4.75	× 60	1951		
4.75 ×	4.75	× 60	1951		
4.75 ×	4.75	× 60	2306		
4.75 in diar.		× 61	2676		
4.62 ×	4.75	× 60	2370		
4.62 ×	4.75	× 60	2826		
4.75 in diar.		× 60	2765		
4.00 ×	4.00	× 78.24	2937	27.6	Heart; unseasoned straight grain; four groups of knots 2 in. 3½ ins. 1¾ ins, 5¾ ins. from end on each face. Crippled and failed through knot 2 ins. from end on low side.
4.03 ×	4.06	× 78.2	3466	28.7	Straight grain; several knots. Failed by bending at knot 30 ins. from one end. Ends square; maximum load 70,500 lbs.
4.03 ×	4.03	× 75	4557	28.2	Straight clear grain; one small knot. Failed at knot 3 ft. 4 ins. from end; crippled, then split open; ends square.
3.95 ×	3.98	× 75	3260	29.3	Grain straight but for frequent knots; failed at a group of knots about 2 ft. from one end by splitting first slightly open and then crippling on one side; it bent afterwards.

RESULTS OF COMPRESSIVE TESTS ON OLD WHITE PINE.

Dimensions in inches. Lengths.			Comp've Strength in lbs. per sq. inch.	Weight per cub. ft. in lbs.	Remarks.
3.5 ×	4.4	× 11.75	1980	27.35	Large knots on all sides about 2 ins. from an end, otherwise in good condition, except shivered at a corner between two knots. Failed by splintering at shivered corner; also crippled at knots.
3.4 ×	4.3	× 11.70	2740	28.10	A large knot appearing on two faces 3 ins. from end; also a slight season crack on one face. Failed by splitting longitudinally along season crack.
3.46 ×	4.32	× 11.75	4470	26.45	Medium knot through corner showing on two faces about 1½ ins. from end; otherwise sound and clear. Failure by crippling at centre.

RESULTS OF COMPRESSIVE TESTS ON OLD WHITE PINE.—Continued.

3.50 ×	4.25 ×	11.74	3850	26.30	Knot on a face 1½ ins. from end, passing to opposite face ½ in. from end; also small deficiency at corner on same end and along one edge; also sapwood. Crippled longitudinally through knot.
3.45 ×	4.39 ×	11.77	4115	25.35	One small pin knot on corner; also shaken by seasoning; also two small injuries on an edge. Burst at the season cracks; afterwards crippled.
3.50 ×	4.41 ×	11.75	2735	25.55	Two large knots at an end on opposite faces 2 ins. from end; also slight season cracks Crippled at knots.
3.47 ×	4.38 ×	11.75	4330	26.50	Clear and nearly straight grained; slightly shaken by season cracks. Crippled 5 ins. from one end.
3.52 ×	4.37 ×	11.75	2625	28.55	A large knot 3 ins. from end passing through from opposite faces; also seasoned somewhat. Crippled through at knot.
3.45 ×	4.25 ×	11.75	4600	23.3	Clear specimen, except deficiency at a corner, partly sapwood; also bad injury (spike hole) in deficient corner. Crippled at centre.
3.45 ×	4.36 ×	11.70	3975	24.5	Two weathered sides; clear; seasoned. Clear crippled at centre.
3.50 ×	4.27 ×	11.70	4695	25.0	One old side; clear shaken by season cracks Crippled at centre.
3.49 ×	4.37 ×	11.75	4230	25.8	Grain clear and straight, large cant-hook hole 1 in. from one end on old narrow side. Failed by crippling at centre.
3.48 ×	4.32 ×	11.73	3910	24.4	Large knot on end; seasoned; grain clear and straight. Failed by crippling at centre.
3.48 ×	4.40 ×	11.74	3830	23.85	Large knot on end; grain clear and straight, season cracks. Failed by splitting longitudinally and crippling slightly at centre.
3.51 ×	4.30 ×	11.60	4525	25.65	One old side; grain clear and straight; piece badly shaken. Crippled at centre.
4.10 ×	4.16 ×	12.00	2923	23.2	Grain clear and straight; season cracks on two old sides; injured by cant. hook on one old side. Crippled at one end and through defect.
4.21 ×	4.19 ×	12.00	2183	23.0	Grain parallel; one small pin knot; season cracks on old side; one small defect on corner 2 ins. from end. Crippled at one end.
4.17 ×	4.18 ×	12.05	2059	25.4	A large knot near centre; badly seasoned on old side; split along seasoning; split from knot. Also crippled.
4.14 ×	4.22 ×	12.00	2840	22.9	Grain clear and straight, seasoning cracks through centre; small defect on old side. Crippled through defects.

RESULTS OF COMPRESSIVE TESTS ON OLD WHITE PINE.—*Continued.*

4.19 ×	4.20 ×	12 00	1716	32.5	A large knot from end to end along one face; another at one end; another at opposite side. Fibre split from knot.
4.18 ×	4.22 ×	12.00	2228	26.3	A large knot from end to end along one face; another at one end. Crippled at knot at centre, and also a splitting away.
4.14 ×	4.18 ×	12.00	2794	23.1	Clear and straight; seasoned on two old sides. Crippled at one end.
4.17 ×	4.19 ×	12.00	1723	25.0	Grain clear and straight, bad season cracks on old side; spike hole 2¼ ins. deep, 2 ins. from one end. Failed at spike hole.
4.21 ×	4.21 ×	12.00	2257	22.3	Grain straight; three fresh sides; one large knot near end; season cracks on old side. Crippled through knot at one end.
4.20 ×	4.22 ×	12.00	2438	23.6	Grain straight; two large knots at opposite ends; season cracks on old side. Crippled on end at a knot.
4.16 ×	4.21 ×	12.00	2569	23.4	Grain straight and parallel, except at one end, where it is curled by vicinity of a knot; otherwise sound. Crippled at sound end.
4.19 ×	4.22 ×	12.00	2030	28.0	Two large knots at one end, otherwise straight and clear; fresh sawn on all sides. Crippled at knots at end.
4.13 ×	4.20 ×	12.00	2686	24.1	Grain straight; three small knots at centre; two old sides injured by several small holes. Fibre split and crippled at small knots.
4.17 ×	4.18 ×	12.00	2180	25.3	Three large knots at centre; grain parallel; full of season cracks on old side; fibre split. Crippled at knots.
4.20 ×	4.21 ×	12.00	1883	24.4	Grain crooked by knots; two large knots near centre; large season crack on one old side. Crippled across centre at knots.
4.21 ×	4 23 ×	12.00	1915	25.0	Four large knots near centre, otherwise clear and straight; one knot at each corner. Crippled across centre at knots.
4.16 ×	4.21 ×	12.00	2512	23.39	Grain straight; three sides fresh sawn; small pin knot; small defect at one end on old side. Crippled at and near small defect.
4.20 ×	4.23 ×	12.00	2277	26.1	A large knot hole at an end; three smaller knots near centre; otherwise sound and straight. Crippled at end aided by knot.
4.18 ×	4.23 ×	12.03	1838	27.2	Two sides fresh sawn; three large knots 2 ins. to 4 ins. from one end; grain twisted; three cant-hook marks; cracks in medullary rays. Failed by splitting from large knot.

RESULTS OF COMPRESSIVE TESTS ON OLD WHITE PINE.—Continued.

4.20 × 4.23 × 12.01	2477	25.0	Three sides fresh sawn; grain not parallel, owing to a knot; one season crack on old side; wood decaying somewhat; several small pin knots. Sheared along season crack, caused by adjacent knot.
4.19 × 4.22 × 12.05	2177	26.4	Three fresh sawn sides; two large knots near centre; one pin knot; grain parallel; very large season cracks. Split along season cracks.
4.20 × 4.25 × 12.04	2387	26.1	Four sides fresh sawn; grain parallel; season cracks are through specimen; one large and two small knots at one end, large one at corner. Crippled at knots.
4.17 × 4.20 × 12.02	2752	24.7	Three sides fresh sawn; grain not parallel; season cracks through body of specimen; slightly decayed on one side; several small pin knots. Sheared on rot line and crippled at knots.
4.21 × 4.23 × 12.02	1797	26.7	All sides fresh sawn; two large knots in body; grain parallel; slight decay; cracks in medullary rays. Crippled through knots.
4.18 × 4.20 × 12.05	1789	25.0	Two sides fresh sawn; grain not quite parallel; large knot at one end; season cracks on two old sides; small knot in body. Crippled through knots.
4.19 × 4.22 × 12.05	2099	24.8	Three sides fresh sawn; grain parallel; season cracks on old side; two small injuries in old side near one end. Crippled through very small knot near one end.
4.21 × 4.22 × 12.01	2251	27.3	Three fresh sides; specimen full of knots, two at one end, one large knot and two small knots in body; bad season crack on old side. Crippled through knot at one end.
4.17 × 4.24 × 12.02	1606	28.0	Four fresh sides; two large knots near centre; two pin knots; grain parallel. Crippled and split along fibre from the knots.
4.18 × 4.20 × 12.0	2033	25.4	Three sides fresh sawn; large knot 4 ins. from end; grain parallel; slight decay. Crippled opposite knot.
4.20 × 4.22 × 12.0	2499	25.9	Four sides fresh sawn; large knot near centre; grain parallel. Crippled opposite knot.
3.82 in diam. × 13.65	5770	30.3	Clear and straight grained. Failed by folding through an injury from cant hook 4½ ins. from end.
3.625 × 4.50 × 40.875	2390	22.4	Grain straight; one old side; free from large knots; failed by bursting open along three lines, which pass through various knots and season cracks.

RESULTS OF COMPRESSIVE TESTS ON OLD WHITE PINE.—*Continued.*

3.75 × 4.31 × 45.25	2970	23.6	Grain straight; one old seasoned side; several knots; failed at one large knot in middle of pillar, which passed through from side to side. Failure by bending across narrow dimension.
3.50 × 4.50 × 45.125	1840	22.6	Grain straight; one old seasoned side; many knots; failed at one large knot in middle of pillar, which passed through from side to side. Failure by bending across narrow dimension.
3.50 × 4.38 × 44.5	2170	21.9	Grain straight; one old side; many small knots; one large knot on old side 15 ins. from one end. Failed by crippling at that knot.
3.73 × 4.35 × 44.5	2650	23.6	Straight grain; fairly clear; some small knots; one old seasoned side. Failed by bending 18 ins. from one end in clear wood across least dimensions.
3.5 × 4.4 × 45	3346	22.8	Grain straight; two old sides; knot at one end; also knot at centre passing through a corner. Failed by direct crippling which started at knot in middle of the piece.
3.5 × 4.4 × 42.5	2082	21.1	Grain nearly straight; one old side; various knots, particularly one near centre passing from corner to corner of section. Failure by bending at this knot on least dimension.
3.5 × 4.45 × 46	2248	21.7	Grain straight; one old side. Failed near centre by bending across least dimension at a knot, which penetrated the heart of piece from one side.
3.83 × 3.83 × 71.3	2862		Two knots on one edge; one large knot at centre, another 12 ins. away; on second face five knots, two near centre, others 12 ins. from ends; grain parallel; centre of tree in corner of specimen, failed by bending at centre knot, induced first by being ¼ in. off centre on top bearing.
3.84 × 3.84 × 72.0	3338	26.06	Bad knot 6 ins. from centre on one face; next face knot 2 ins. from end; grain about parallel; many smaller knots; centre of tree on same corner as large knot. Failed by bending at large knot.

RESULTS OF COMPRESSIVE TESTS ON

NEW SPRUCE (B.C.)

Dimensions in inches. Lengths.	Compress've Strength in lbs. per sq. inch.	Weight in lbs. per cub. ft.	Remarks.
4.72 × 2.343 × 1.94	3415		
4.77 × 2.25 × 1.9	2941		
4.75 × 2.375 × 1.875	3020		
4.72 × 2.25 × 1.875	3465		
4.78 × 2.25 × 1.97	3256		
4.75 × 2.25 × 1.94	3118		
4.75 × 2.312 × 1.88	3009		
4.72 × 2.22 × 1.9	3179		
3.75 × 2.34 × 1.62	3854		
4.812 × 2.312 × 1.94	3210		
4.375 × 1.875 × 2	4440		
4.75 × 2.25 × 2.50	3321		
4.73 × 4.73 × 3.9	3451		
3.67 × 3.67 × 3.64	5590		Failed by crippling.
4.75 × 4.75 × 4.0	3325		
4.75 × 4.75 × 4	2838		
4.812 × 4.812 × 4	2986		
4.65 × 4.65 × 5.20	4540		
3.00 × 2.875 × 6.50	7566		Clear and straight.
3.00 × 3.125 × 6.00	6036		
4.7 × 4.7 × 7.75	4299	29.80	Four pin knots; ends not quite parallel.
3.125 × 2.875 × 7.25	6312		
4.687 × 4.687 × 8.66	5305	29.80	Clear and sound; cracks along medullary rays.
4.75 × 4.75 × 11.5	4656		Clear and straight.
4.2 × 3.8 × 11.5	4806	25.9	Crippled at centre.
4.0 × 4.04 × 11.75	3898	33.8	Straight grained. Crippled at large knot on edge near centre.
4.10 × 4.10 × 12.55	4451	28.3	Clear and straight grained; slight axe-cut on one face 3 ins. from end. Failed by crippling at axe cut.
3.75 × 3.75 × 12.05	4907	29.5	Crippled at a bunch of five knots.
4.72 × 4.72 × 14.09	4063	30.2	Five large knots and one large season crack.
4.75 in diar. × 14.	3328		
3.33 × 4.18 × 14.97	4382	33.9	Clear and straight. Failed by crippling near one end.
4.35 × 4.32 × 20.55	3757	29.6	Failed by crippling.
4.35 × 4.45 × 20.6	3540	27.1	Knot near one end. Failed in centre.
4.41 × 4.45 × 20.6	3850	29.9	Clear.
2.5 × 3.42 × 27.5	3390	26.3	Clear and straight grained, but heavy season crack from side to side. Failed by bulging on season crack and then bending.
3.48 × 3.50 × 32.25	4384		Grain not straight; heavy knot through centre; also ends not square. Burst apart along centre.
2.75 × 4.05 × 41.0	3070	28.3	Straight grained. Failed at large knot 3 ins. from end by crippling.
2.75 × 4.02 × 40.95	3086	28.4	Straight grained; eight large knots. Failed by bending at two knots 19 ins. from one end concave to high side.

RESULTS OF COMPRESSIVE TESTS ON NEW SPRUCE (B.C.)—*Continued.*

4.35 × 4.50 × 20.55	3584	27.4	Grain clear and parallel. Crippled at centre.
4.08 × 4.35 × 22.97	3909	27.5	Grain crinkled near one end. Failed there.
4.18 × 4.35 × 22.95	3271	27.7	Clear; straight; no knots. Failed at one end.
4.29 × 4.35 × 22.96	3617	25.4	Grain not quite parallel; knot near centre of one side at which piece failed.
4.20 × 4.35 × 22.95	2834	28.2	Grain not parallel. Failed by longitudinal shear, which passed through a knot.
4.25 × 4.40 × 22.9	3774	26.1	Failed at a knot near centre of one side.
4.24 × 4.34 × 22.94	2973	25.1	Failed by longitudinal shear.
4.12 × 4.35 × 23.00	3560	27.2	Failed at a knot.
4.10 × 4.41 × 23.00	3680	25.7	Grain parallel. Failed by crippling at a knot 6 ins. from one end.
4.25 × 4.40 × 23.0	3382	27.9	One season crack, did not affect the failure which was by crippling.
4.10 × 4.40 × 23.05	3550	26.4	Knot near one end. Crippled in body of piece at a distance from the knot.
4.09 × 4.35 × 23.06	4229	25.6	Grain clear and parallel. Crippled on one side.
2.97 × 4.0 × 15.1	4908	26.7	Clear and straight grained. Crippled two from end.
3.33 × 4.1 × 15.04	3370	26.4	Straight grained; large knot on middle of side. Failed near one end in clear wood.
4.72 in diar. × 15.0	3430	30.86	Four deep medullary weathering cracks; a mass of knots at lower end; small pin knots at centre; ends not quite parallel. Crippled at lower end at knots.
2.6 × 4.1 × 18.5	5253	24.1	Clear and straight grain; failed by crippling and bending 6 ins. from one end.
4.75 in diar. × 60	1862		
4.75 " × 60	2708		
4.75 × 4.75 × 60	2351		
4.75 × 4.75 × 60	2275		
4.75 × 4.75 × 60	3104		
4.75 × 4.75 × 60	2660		
4.75 × 4.75 × 60	2351		
4.75 × 4.75 × 60	2306		
4.75 × 4.75 × 60	2661		
4.62 × 4.63 × 60	2431		
4.62 × 4.75 × 60	2416		
4.62 × 4.62 × 60	2420		
4.75 in diam. × 60	2483		
4.75 " × 61	2483		
4.75 " × 61	3215		

RESULTS OF COMPRESSIVE TESTS ON
OLD SPRUCE.

Dimensions in inches.			Lengths.	Compr've Strength in lbs. per sq. inch.	Weight in lbs. per cub. ft.	Remarks.
2.54	×	3.15	× 5.95	4375	28.4	Clear wood, straight grained; ends out of square; bent over.
2.12	×	2.97	× 10.12	4508	28.4	Clear wood, straight grained; ends out of square; bent over.
2.42	×	2.45	× 10.95	4367	27.9	Clear wood, straight grained; failed by bending; worm eaten.
2.50	×	3.20	× 11.25	3862	28.4	Clear wood, straight grained; ends out of square; bent over.
2.18	×	2.18	× 14.00	4842	27.9	Clear wood, straight grained; failed by bending; worm eaten.
2.17	×	2.18	× 13.40	4714	27.9	Clear wood, straight grained; failed by bending; worm eaten.
3.20	×	3.22	× 13.40	5825		Clear; straight grained; crippled at centre.
3.20	×	3.21	× 13.28	5696		Clear; straight grained; crippled at end at a previous injury on surface.
3.17	×	3.21	× 13.62	4900		Straight grained; knot at centre. Crippled at knot.
3.20	×	3.20	× 13.43	5273		Straight grained; knot on corner at centre. Failed at knot.
2.80	×	3.35	× 13.30	5139		Heavy knot through edge near centre. Crippled at knot.
2.80	×	3.34	× 12.50	4818		Straight grained; knots near each end. Crippled and burst through large knot.
2.18	×	2.18	× 16.00	4337	27.9	Clear wood; straight grained. Failed by bending; worm eaten.
3.53	×	3.56	× 14.60	6329		Clear and straight grained. Crippled near end through a small injury like a nail hole.
2.60	×	2.63	× 15.45	7930		Clear; straight grained. Crippled 5 ins. from end.
2.60	×	2.75	× 16.25	3664		One small knot, but badly out of parallel. Failed at knot.
2.66	×	2.65	× 15.57	6809		Straight grained; one small knot near end. Crippled first near centre through cant hook holes
2.80	×	3.37	× 27.05	5116		Straight grain; knot 12 ins. from end. Crippled at knot.
2.80	×	3.35	× 26.26	5096		Straight grain; knot 10 ins. from end. Crippled at a knot.
2.62	×	2.75	× 17.72	5625		Clear, but grain very much out of parallel, as much as 3 ins. in 18 ins. Burst apart by shearing of unsupported fibre.

TENSILE STRENGTH.

The experiments were especially directed to the comparison of the tensile strength and stiffness of portions of the same stick, in different positions relatively to the heart.

In designing the form of the test-piece, it was of importance to make the head of such a depth as would prevent the central portions from being pulled through the head by shearing along the surface BC, and it was also necessary that the depth should not be inconveniently great. Wedge shaped holders (Fig. H) were adopted which would grip the

Fig. E.

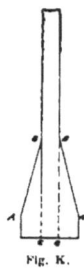

Fig. K.

specimen along the faces AB. This form of holder was intended to increase the resistance to shear which is always much less than the tensile strength. As the tension on the test-piece increases, so also does the normal pressure upon the faces AB, Fig. K, and, therefore, so also does the resistance to shear along the surface BC. At first, the faces of the holders in contact with the specimen were left rough, but it was found that the roughness prevented the specimen from sliding in far enough to be gripped along the whole of the face AB, so that the bearing surface was practically limited to a comparatively small area near the top of the head. Thus it often happened that the specimen still failed by shearing along the surface BC. This difficulty was obviated by planing the faces of the holders.

The test-pieces were prepared from the uninjured portions of the beams, which had already been fractured transversely. The extensions of a length of ten inches of the specimen under gradually increased loads were measured by means of Unwin's extensometer until the total extension exceeded about one-eightieth of an inch. After this the extensometer was removed, and in many cases additional extension readings, up to the point of fracture, of a length of sixteen inches of the specimen, were measured by means of a steel rule and indicator clamped to the specimen at points 16-inches apart and allowed to slide over one another.

The results obtained are given in the following Tables and an examination of these will show :—

1st. That the increments of extension up to the point of fracture are almost directly proportional to the increments of load ;

2nd. That the presence of knots is most detrimental both to the strength and to the stiffness, inasmuch as they practically diminish the effective sectional area, and also produce a curvature in the grain ;

3rd. That wood near the heart possesses much less strength and much less stiffness than that more distant from the heart ;

4th. That the strength and stiffness are also dependent upon the proportion of summer to spring growth;

5th. That irregularity of readings, both with the extensometer and with the rule, are chiefly due to the presence of a knot, or to curly or oblique grain caused by a knot.

Again, some of the Tables give the effects on various specimens, of alternately loading them and relieving them from their load, and from the experiments carried out up to date the following inferences may perhaps be drawn:—

If the specimen is clear, free from knots, and straight in the grain, and if no interval of rest is allowed, then for any given range of loads:

(a) The total extension is greatest during the first loading;

(b) The extensions due to the successive loadings continually diminish, tending to a minimum limit, so that the co-efficients of elasticity increase, and therefore so also does the stiffness;

(c) By the successive unloadings a set is produced, which continually increases, but at a diminishing rate, and which tends to a maximum limit;

(d) When the specimen is allowed an interval of rest under the minimum load, the first total extension, when the loading is resumed, is greater than at the commencement, but continually diminishes, tending to a minimum limit, which possibly coincides with the maximum limit reached previous to the interval of rest.

So also, after the interval of rest, when the first set produced the specimen is from load, is greater than that previously produced, but gradually diminishes, in the succeeding releases from load, tending probably to a minimum limit coinciding with the maximum limit reached before the interval of rest.

These inferences are also in accord with similar experiments carried out by Mr. Kerry, B.A.Sc.

Special attention may be directed to the test of specimen 4, beam XXI. This specimen failed simultaneously at two sections, the wood seeming to be very brittle, and the character of the failure pointed to some inherent weakness in the timber itself. After a microscopic examination of the fractured sections, Professor Penhallow described the fractures as being "very regular and devoid of any fibrous character, having the "exact appearance of a piece of glass. The lines of fracture followed "the variations in thickness of structure longitudinally and trans-"versely with great regularity. The peculiar brittleness can only be "referred to some local molecular condition of unknown origin, possibly "to a deficiency in the element of water."

The simultaneous failure at two sections of specimens 2 and 8 from White Pine beam XLVIII may probably be referred to a similar cause, and, as Professor Penhallow says, adequate explanations of such failures are still to be sought.

In the Tables the extensometer measurements are given in hundred-thousandths of an inch, and the rule measurements in hundredths of an inch.

With each table a diagrammatic section is also given, showing the part of the stick from which the several specimens have been taken.

DIAGRAMETIC SECTIONS FOR TENSION SPECIMENS.

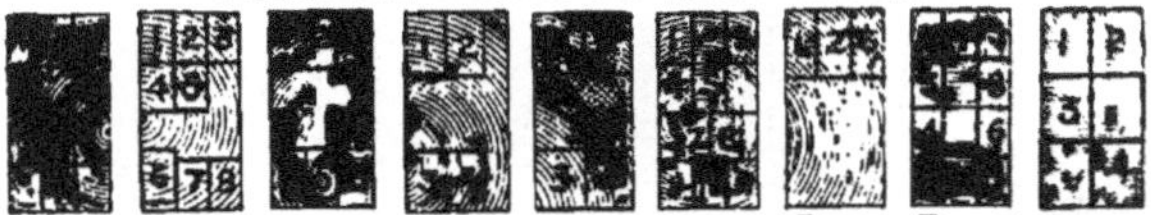

Fig. 118 Fig 119. Fig 120. Fig 121. Fig 122. Fig 123 Fig 124 Fig 125 Fig 126

Fig 127 Fig 128. Fig 129 Fig 130 Fig 131. Fig. 1 Fig. 1 Fig. 132 Fig. 133

Results of tension tests on specimens 1 to 9 cut out of Douglas Fir Beam IX, and of repeatedly loading a specimen cut out of the same Beam. (Fig 118.)

Loads in lbs.	Readings taken by Extensometer. Specimen.							
	1 Forward.	2 Forward.	3 Forward.	4 Forward.	6 Forward.	6 Forward.	7 Forward.	9 Forward.
100	0	0	0	0	0	....	0	0
200	81	79	65	92	80	....	50	82
400	229	227	194	261	240	259	162	252
600	372	379	318	430	393	....	293	421
800	509	527	435	579	549	564	403	579
1,000	614	673	547	737	702	....	520	736
1,200	779	818	664	870	852	863	637	890
1,400	914	960	784	1060	1004	1004	752	1047
1,600	1049	1097	894	1226	....	1183	869	1200
1,800	1185	1241	1008	1395	....	....	984	....
2,000	1323	....	1124	....	....	....	1098	....
Total breaking weight in lbs.	9270	6290	10,580	8820	6390	....	10,114	6348
Break'g weight in lbs. per sq. in.								
Coefficient of elasticity in lbs.								

Results of tension tests on specimens 1 to 7 cut out of Douglas Fir Beam X.

Loads in lbs.	1 Extr.	Rule	3 Extr.	Rule	5 Extr.	Rule	4 Extr.	Rule	4 Extr.	Rule	4 Extr.	Rule	7 Extr.	Rule	6 Extr.	Rule
100	0		0		0		0		0		0		0		0	
200	89	0	62	0	63	0	68	0	86	0	99	0	98	0	65	0
300	221	3	157	5	136	6	198	2	260	2	241	2	238	5	208	6
400	333	12	277	10	251	10	319	8	432	16	461	10	416	12	347	11
500	481	19	397	15	367	16	463	13	613	24	633	17	728	20	495	17
600	614	19	515	20	484	20	679	19	788	32	807	33	578	27	620	23
700	743	29	636	25	592	25	879	23	943	40	970	37	900	34	755	29
800	878	37	755	30	712	31	1,049	29	1,135		1,132	45	1,057	41	890	35
1000	978	47	877	35	819	33	1,175	36	1,303			52	1,270	48	1025	41
1200	1,165	55	968	40	953	40	1,573	44	1,50			60	1,429	55	1139	47
1400	1,258	62	1,121	45	1,068	48		52				73		62	1297	52
1600		69		50		60		67				82		70		60
1800		79		55		68		71						79		66
2000		89		60		73		76						89		71
2500		94		65		75		83								
3000		100		70				90								
Total breaking weight in lbs.	9,500		9,500		9,240		10,850		4,820		6,260		7,965		7,500	
Br'k'g weight in lbs. per sq. in.	11,902		13,532		13,368		15,116		6,914		9,664		11,326		11,623	
Co-effic't of elasticity in lbs.	1,830,500		2,364,300		2,510,650		1,710,330		1,650,400		1,665,400		1,686,150		2,009,000	
Time of test in minutes.	18		25		30½		17		18		12		26			

Results of repeatedly loading tension specimens 2 and 5 cut out of Douglas Fir Beam X. (Fig. 119.)

Loads in lbs.	2 Extensometer									Rate	5 Extensometer									Rate
	For-ward turn	Re-turn	For-ward turn	Re-turn	For-ward turn	Re-turn	For-ward turn	Re-turn	For-ward turn	Re-turn	For-ward turn	Re-turn	For-ward turn	Re-turn	For-ward turn	Re-turn	For-ward turn	Re-turn	For-ward turn	Re-turn
100	0	3	3	1	5	8	8	0	0	29	29	41	41	63	63	63
200	26	26	56	70	62	13	78	67	51	91	85	103	98	117	109	125	119
300	58	48	62															12
400	176	178	169	180	191				167	215	213	232	222	247	236	253	234	18
500	294		316															21
600	418	423	429	439	445	450	467	452	470	400	283	436	447	480	453	489	490	500	472	23
800	540	427	448		461	461					403									25
1000	675	680	701	686	713	701	720	706	723	713	526	693	690	721	700	732	711	743	717	30
1200	791	683	704								632									35
1400	919	930									775	927	933	955	943	966	954	976	961	40
1600	1049	1068	1073	1080	1087	1087	1091	1091	1095	0	900									45
1800										6	1020									50
2000										12	1150	1150	1176	1176	1184	1184	1199	1199	1203	60
2500										18										70
3000										23
3300										29
4000										35
4500										40
5000										45
5500										46
6000										51

Total breaking weight in lbs.	7,000	7,500
Break'g weight in lbs p. sq. in.	10,145	10,757
Co-efficient of elasti'y in lbs.	2,321,600	2,334,850
Time of test, in minutes	49	45

Results of tension tests on specimens 1, 2, 5 cut out of Beam X, and of repeatedly loading specimens 3, 4, 6 cut out of same Beam. (Fig. 119.)

Specimen Readings taken by Extensometer.

Loads in lbs.	1 Extr. For- ward	1 For- ward	2 Exten- som. For- ward	3 For- ward	3 Re- turn.	3 For- ward	3 Re- turn.	3 For- ward	3 Re- turn.	4 For- ward	4 Re- turn.	4 For- ward	4 Re- turn.	4 For- ward	4 Re- turn.	4 For- ward	5 For- ward	6 For- ward	6 Re- turn.	6 For- ward	6 Re- turn.	6 For- ward
100	0		0	0						6							0	0				
200	62		52	65						78							72	71				
400	172		154	213	259	259	262	262		194	201	201	212	212	221	225	220	196	213	213	218	218
600	284		256	356	310	327	331	321		326	481	492	490	500	506		364	329				
800	403		365	497	527	527	531	521		469		195		506	303		507	467	468	484	490	489
1,000	523		468	631	801	801	816	803		612	737	760	761	763	768	774	647	609				
1,200	646		576	771						750					772		786	689	763	759	767	762
1,400	769		678	907	1036	1070	1070	1074		887							924	730				
1,600	892		775	1030						1019	1027	1027	1033	1039	1039	1041		890				
1,800	1013		873					1210								1193		1029	1029	1031	1031	1037	1172
2,000	1136		971														1189					
2,200	Failed at section where grain was		1044																			
2,500	curly, due to proximity of a knot.									This test-piece was cut out from							Failed at two small knots.					
3,000			13							near the heart.												
3,500			18																			
4,000			23																			
4,500			29																			
5,000			31																			
5,500			35																			
6,000			40																			
6,500			43																			
7,000			47																			
7,500			51																			
8,000			57																			
8,500			61																			
9,000			66																			
9,500			70																			
10,000			74																			
10,500			78																			
11,000			83																			
11,500			89																			
12,000																						
Total breaking weight in lbs.	7460		12413	7228						7340							6680	8424				
Break'g weight in lbs. p. sq. in.	10,376		17,492	10,191						10,279							992	11,535				
Co-efficient of elasticity in lbs.	2,308,650		2,816,900	2,021,250						2,036,800							2,134,150	1,973,150				

Results of tension tests on specimens cut out of Douglas Fir Beam X, and of repeatedly loading another specimen cut out of same Beam (Fig. 119).

Readings taken by Extensometer.

Loads in lbs.	Specimen 7							Specimen 8		Specimen 7			
	For-ward	Return	For-ward	Return	For-ward	Return	For-ward	For-ward	For-ward	Return	For-ward	Return	For-ward
100	0	...	...	...	...	...	...	0	0	...	...	...	...
200	58	...	...	...	...	...	...	69	78	...	...	...	...
400	174	172	172	156	111	179	179	214	193	222	222	228	228
600	299	110	112	413	416	118	417	341	316	158	453	463	228
800	417	...	...	...	...	...	...	468	430	...	...	...	...
1000	534	636	636	639	660	663	661	602	545	680	680	689	459
1200	651	...	...	...	...	...	...	731	760	...	...	...	683
1400	776	...	...	...	...	...	...	860	875	...	...	...	...
1600	898	...	...	...	...	...	...	983	890	...	...	...	...
1800	1019	1019	1023	1023	1023	1027	1029	1121	1003	1015	1017	1017	1020
							1153		1120	1120	...	...	113%
200*	2030							700	9270				
Total break'g wt. in lbs.	9742							1,1140	13,071				
Brk'g weight in lbs per sq. in	2,296,550							2,223,150	2,9,6530				
Co-efficient of elasticity in lbs.													

* After this, the 4th series of readings, the test-piece was allowed to rest for a period of 2 hours. On resuming the testing the reading was 500lb.
Note.—In test-pieces 7, 8 and 9, the grain was somewhat oblique to the direction of the axis.

Results of tension tests on specimens 1 to 6 cut out of Douglas Fir Beam XII, and of repeatedly loading specimen 3 cut out of same Beam. (Fig. 120).

Loads in lbs.	Readings taken by								3 Extensometers							
	1 Extr.	2 Extr.	3 Extr.	4 Extr.	Rate	5 Extr.	6 Extr.	1 Extr.	2 Extr.	For- ward	Ret. turn.	For- ward turn.	Re- turn.	For- ward turn.	Re- turn.	For- For- ward ward
100	0	0	0	0		0	0	0	0	0	0	57	57	70	70	82
200	79	82	69	39		66	73	79	75	54						
300	212	232	192	142		170	220	228	211	172	232	225	241	237	259	246
400	383	382	313	263		284	317	384	363	299						
500	545	548	442	387		390	517	540	525	419	467	441	478	465	489	475
600	704	720	600	505		509	666	702	670	539						
1,000	856	875	759	629		613	813	856	817	665	692	687	708	697	715	705
1,200	1065	1032	907	749		722	953	1009	953	792						
1,400	1153	1129	1030	869		848	1107	1118	1100	914						
1,600			1214	986		919				1032	1032	1040	1040	1042	1042	1051
1,800				1106		1063										1178
2,000					0											
2,200					5											
2,500					10											
2,800					15											
3,000					20											
3,200					25											
3,500					30											
4,000					35											
4,500					40											
5,000					45											
5,500					50											
6,000					55											
6,500					60											
7,000					66											
7,500					73											
8,000					81											
8,500					90											
9,000					100											
9,500					110											
10,000																
10,500																
11,000																
Total break'g w'ght in lbs.	10,620	10,760	10,700	11,120		9,900	5,310	10,220	9,300	10,420						14,640
Br'k'g weight in lbs. p. sq. in.	14,886	15,327	15,040	15,655		13,909	7,823	14,660	13,066	14,640						
Coeffic'nt of elast'y in lbs	1,791,800	1,805,050	2,001,650	2,307,200		2,486,700	1,920,900	1,934,100	1,816,500	1,960,450						
Time of test in minutes	10	10	10	16		12	9	8	9	41						

Notes in table:
- Column 4 Extr.: "In this test-piece the central portion was pulled through the head, so that its tensile strength exceeded the breaking weight."
- Column 5 Extr.: "This test-piece commenced to fail at a small knot."
- Column 6 Extr.: "This test-piece failed at a section where the grain was curly, from proximity to a knot."
- Right side: "In this test-piece the central portion was pulled through the head, so that its tensile strength exceeded the br'kg weight."

86

Results of tension-tests on specimens cut out of Beam XIII, and of repeatedly loading other specimens cut out of the same Beam (Fig. 121).

Readings taken by

Loads in lbs.	Extensometer 1							Rule	Extensometer									
									2	3	4	1		2	3			
	Forward	Re-turn	For-ward	Re-turn	For-ward	Re-turn	For-ward	For-ward	Forward	Forward	Forward	Forward	Forward	Forward	Forward	Forward	Re-turn	For-ward	Re-turn	For-ward
100	0	...	...	...	...	...	...	...	0	0	0	0	0	0	...	...	...	...
200	64	...	...	...	...	...	...	...	70	119	64	101	62	85	67	...	...	...
300	172	172	180	182	198	198	...	...	199	306	215	216	181	287	191	259	239	253
400	180	...	...	...	...	199	199	202	337	491	338	120	300	464	315	...	...	...
600	295	...	416	418	135	434	126	135	130	472	613	465	572	418	620	447	390	455	506	139
800	106	409	403	...	...	...	...	...	610	839	591	723	534	786	573	...	...	...
1,000	516	623	627	631	617	659	661	653	664	656	742	1015	728	816	652	950	766	730	239	745	751
1,200	620	...	...	...	...	...	...	...	872	1179	844	1030	772	1110	831	...	...	...
1,400	721	...	...	...	...	...	...	...	1006		962	1185	891		957	...	...	...
1,600	832	939	951	951	975	973	981	981	983	953	1140		1090		1012		1078	1078	1102	1102	118
1,800	939	...	...	...	...	...	...	...			1210		1132						
2,000	...	...	...	...	...	...	983	983	1092	0									
2,500	...								5									
3,000	...								10									
3,500	...								15									
4,000	...								20									
4,500	...								25									
5,000	...								30									
5,500	...								34									
6,000	...								39									
Total break'g w'ght in lbs.	7,520								9840	5140	8720	7490	11,620	1370	9320				
lbs/kg weight in lbs. per sq. in.	10,638								13,946	7322	12,337	10,191	15,271	6278	13,721				
Co-efficient of elasticity in lbs.	2,609,400								2,108,500	1,631,700	2,263,100	1,684,900	2,359,150	1,092,900	2,223,700				

* After this, the 4th series of readings, the test-piece was allowed to rest for 2½ hours. On resuming the testing, the reading was 1095.

87

Results of repeatedly subjecting to tensile stress a specimen cut out of Beam XV. (Fig. 122.)

Specimen 1.

Readings taken by Extensometer.

Loads in lbs.	Forward.	Re-turn.	For-ward.	Re-turn.	For-ward.	Re-turn.	For-ward.	Re-turn.	For-ward.	Re-turn.	For-ward.	Re-turn.	For-ward.	Re-turn.	For-ward.	Lobs.
100	0	20	20	20	20	22	22	22	22	44	41	43	43	49	49	0
200	79															6
400	229	229	216	231	229	232	230	230		237	229	238	230	240	237	12
600	379															18
800	522	500	495	509	494	511	492	510	519	542	529	546	530	547	536	23
1,000	659															29
1,200	790	790	771	784	772	790	780	788	797	816	807	821	809	821	813	35
1,400	926															40
1,600	1059															46
1,800	1186	1186	1189	1181	1183	1183	1184	1191	1215	1215	1219	1219	1219	1219	1220 1358	51
2,000																55
2,250																57
2,500																63
3,000																69
3,500																75
4,000																76
4,500																81
5,000																90
5,500																
6,000																
6,500																
7,000																
7,500																
8,000																
8,500																
9,000																
9,500																

Total breaking load in lbs. 10000
Breaking load in lbs. p. sq.in. 14,474
Co-efficient of elasticity in lbs. 2,092,600

* After this 8th series of readings the test-piece was allowed to rest for a period of 16 hours under the load of 100 lbs. On resuming the testing the initial reading was found to be unchanged.

Loads in lbs.	Extensometer 2			Extensometer 3			Extensometer 1			Rule		Extensometer 4	
	For-ward	Re-turn	Extr. For-ward	For-ward	Re-turn	Extr. For-ward	For-ward	Re-turn	Extr. For-ward	For-ward	Extr. For-ward		Extr. For-ward
100	0	30	39	0	39	39	0	21	21	0	35		0
200	62			63			50				178		69
400	187	224	299	199	239	230	157	173	136	70	311	6	189
600	313			337			263		361	211	441	10	306
800	439	453	479	457	509	495	366	395		234	572	16	418
1,000	565	600		476	611		473		573	499	703	21	530
1,200	694	707	725	746	764	756	578	599		341	833	26	642
1,400	820	843		869			684			601	703	32	732
1,600	949			992			785		875	738	962	37	847
1,800	1073	1086	1084	1090	1130	1132	844	899		969	1091	43	936
2,000	1201	1201		1217		1288	965			999	1220	48	1073
2,200							1088			1129		53	
2,500												61	
												66	
												69	
												76	
												81	
												88	
												94	
												102	

Total breaking weight in lbs.	10950	7420	7220	10240	11000	8115
Breaking weight in lbs. p. sq. in.	15346	10,610	11,117	15,004	15,619	11,686
Co-efficient of elasticity in lbs.	2,206,250	2,144,850	2,181,550	2,231,300	2,173,350	2,626,200

Results of repeatedly subjecting to tensile stress a specimen cut out of Beam XV. Fig. 122.

Specimen 4.

Readings taken by Extensometer.

Loads in lbs.	Forward.	Re-turn.	For-ward	Re-turn.	For-ward	Re-turn.	For-ward	Re-turn.	For-ward	Re-turn.	For-ward	Re-turn.	For-ward	Re-turn.	For-ward	Re-turn.	For-ward	Re-turn.	For-ward	itude.
100	0	38	58	69	69	75	75	75	77	0	9	9	16	16						
200	51									43										
300	163	190	189	202	197	212	203	212	214	125	121	111	114	108	114	111	112	112	112	
400	245									212										
600	338	369	361	365	368	393	374	393	394	301	310	285	299	281	295	287	294	288	288	
800	430									387										
1,000	524	554	545	567	560	574	571	556	576	475	477	461	468	369	463	459	461	458	459	
1,200	620									563										
1,400	713									630										
1,600	809	825	826	826	834	833	833	834	834	738	742	723	723	720	720	716	716	716	716	
1,800	904									825	825									
2,000	1000	1000																		
2,200																			803	
2,400																			893	
2,600																			982	
2,800																			1072	
3,000																			1161	0
3,500																				7
4,000																				11
4,500																				13
5,000																				17
5,500																				17
6,000																				21
6,500																				25
7,000																				29
7,500																				31

Total breaking weight in lbs.		12500
Break'g w'ht in lbs. per sq. in.		19001
Co-efficient of elasticity in lbs.		3,141,900

8,000
8,500
9,000
9,500
10,000
10,500
11,000
11,500
12,000
12,500

* After this the 10th series of readings, the test-piece was allowed to rest entirely free from load for a period of 46 hours.

Results of tension tests on specimens 1 to 11 cut out of Douglas Fir Beam XVII. (Fig. 123.)

Loads in lbs.	1 Extr.	Ital.	3 Extr.	Ital.	4 Extr.	Ital.	8 Extr.	Ital.	9 Extr.	Ital.	11 Extr.	Ital.	7 Extr.	Ital.	5 Extr.	Ital.
100	0		0		0								0		0	
200	61		36		38											
400	185		165		177		95		71		101		91		93	
600	286		278		301		289		210		266		299		240	
800	408		391		425		471		344		419		496		393	
900									481							
1,000	511		505		546		655		612		560		680		550	
1,200	618		620		669		843		745		708		880		699	
1,400	735		734		787		1,037	0	877	0	848		1,073		854	
1,600	834		845		909						996				1,006	0
1,800	955	0	960	0	1,026	0					1,144	0	1,271	0	1,159	18
2,000	1,060		1,023	5	1,153	5		12		3	1,285	3			1,313	25
2,200		5	1,185	14	1,279	10		20		9		9		9		
2,500		11		18		16		30		17		17		19		
3,000		18		19		22				13						
3,500		22		24		28				18						
4,000		28		29		32				23						
4,500		32		33		37				32						
5,000				38		43				38						
5,500				42												
6,000				48												
6,500				53												
7,000				58												
7,500																
8,000																
Total breaking weight in lbs.	5,500		8,150		6,500		3,200		5,160		3,000		2,320		3,000	
Break'g weight in lbs. per sq. in.	7,755		11,631		8,933		4,230		7,035		4,320		4,089		4,040	
Co-efficient of elasticity in lbs.	2,578,350		2,518,200		2,224,750		1,377,000		2,036,200		1,978,150		1,426,000		2,264,500	
Time of test in minutes.	27		18		23		14		13		23		18		15	

Remarks:
- Specimen 1: Failure largely due to shear.
- Specimen 8: Failed at a kno[t]
- Specimen 9: Failed at a knot ½ in. in diameter.
- Specimen 11: Oblique grain.
- Specimen 7: Oblique grain.
- Specimen 5: Failed at a knot.

Results of tension tests on specimens 1 to 3 cut out of Douglas Fir Beam XIX. (Fig. 124.)

Readings taken by

Loads in lbs.	1 Extr.	Intl.	2 Extr.	Intl.	3 Extr.	Intl.	1 Extr.	Intl.	2 Extr.	Intl.	3 Extr.	Intl.
100	0	0	0	0	0	0	0	0	0	0	0	0
200	50	5	50	5	58	7	57	5	50	10	61	10
300	190	11	183	12	212	13	190	11	153	16	190	16
400	329	19	290	16	313	19	296	18	279	21	339	21
600	418	25	384	21	423	24	422	26	400	26	478	29
800	589	35	520	26	530	30	541	31	520	31	620	36
1,000	741	45	619	31	661	45	663	29	642	41	760	41
1,200	864	54	730	36	790	39	782	31	766	51	898	51
1,400	1,008	45	863	40	902	43	901		890	59	1,041	59
1,600	1,123	54	970		1,034		1,033		1,004	70	1,142	70
1,800			1,090				1,151		1,310		1,240	
2,000												
2,200												
2,500												
3,000							1,308					10
3,500												16
4,000												22
4,500												25
5,000												33
5,500												39
6,000							1,000	11				44
6,500							1,101	41				52

In this test-piece the central portion was milled through the bowl, so that its tensile strength exceeded the breaking weight.

In this test-piece the central portion was milled through the bowl, so that its tensile strength exceeded the breaking weight.

Total Breaking Weight in lbs.	Break'g w't in lbs. per sq. in.	Co-efficient of elasticity in lbs.	Time of test in minutes.
7,000			
7,500			
8,000			
8,500			
9,000			
9,500			
10,000			
10,500			
11,000			
11,500			
12,000			
11,140	15,543	2,082,700	......
12,600	17,199	2,407,950	......
10,700	14,581	2,320,950	......
11,520	16,960	2,451,150	18
12,480	18,956	2,450,600	15
9,500 {76, 89, 93, 107, 112, 121}	14,210	2,279,350	19
12,300 {45,50,53,56,59,65,70,75,82,85,90}	16,805	2,687,000	28
8,200 {56,66,73}	11,725	2,197,750	22

Results of tension tests on specimens cut out of Douglas Fir Beam XX, (Fig. 125,) and of the repeated loading of other specimens cut out of same Beam:—

Loads in lbs.	Extensometer.				Readings taken by									
	2				3		4		6		8		9	
	Forward.	Return.	Forward.	Forward.	Extr.	Italc.	Extr.	Italc.	Extr.	Italc.	Extr.	Italc.	Extr.	Italc.
100	0		70		0	0	0	0	0	0	0	0	0	0
200	220	70	243		59	71			95		69		90	
400	619	252	641		180	240			232		218		235	
600	987	530	1,005		263	375			364		249		376	
800	1,366	991	1,285		429	510			516		485		521	
1,000		1,366	1,773		563	650			680		623		672	
1,200				5	688	791			827		759		840	
1,400				10	835	931			970		900		1,011	
1,600				15	942	1,075			1,095		1,038		1,117	
1,800				20	1,063	1,210			1,222		1,174		1,201	
2,000				28	1,182	1,369	0	2	1,357	0	1,310	0	1,501	0
2,200				32	1,310									
2,400				11										
2,500				19										
2,600				52							4			5
2,700				61						4				
2,800							3	10	10	10	10	10	12	12
3,000														
3,200														

Extr. Forward	Extr. Return	Extr. Forward	Extr. Italc.
		0	0
			70
		116	194
		233	324
		361	455
0	22	473	589
306		603	721
	321		851
597	612	719	976
		645	1,102
692		960	1,232
		1,080	
		1,205	
		1,320	5
1,190	1,190	1,440	
		1,570	7

3,100										
3,500			15		15			18		15
3,700						13				
3,800			21		20	20	15	21		20
4,000										22
4,200			27		26	26	20	30		30
4,500			31		30	30	25	36		35
4,700			38		39	32	30			40
5,000			41		45	39	35	40		46
5,200			50			44	42	50		52
5,500					50	51		52		56
5,700			57				50	56		61
6,000					55	60	55	61		
6,200					65	63	61			
6,500					70	70	70			60
6,700					80	80				66
7,000						86				70
7,200										77
7,500										
7,700										
8,000										
8,500										
9,000										

(Middle column "Failed at a pin knot.")

Total breaking weight in lbs.	3,190		7,700	9,360	8,840	8,500	7,500	9,760		9,000
Break'g wgt. in lbs. per sq. in.	4,631		10,783	13,265	12,133	12,021	10,610	14,171		12,710
Co-efficient of elasticity in lbs.	1,769,560		2,226,150	2,008,450	1,921,350	2,072,850	1,787,500	2,440,330		1,985,530
Time of test in minutes.	15			20	18	16	20	14		22

Results of tension tests on specimens cut out of Douglas Fir Beam XXI., and of the repeated loading of another specimen cut out of same Beam. (Fig. 126.)

Loads in lbs.	Readings taken by 1				2		3		4	
	Extensometer.			Rule	Extr.	Rule	Extr.	Rule	Extr.	Rule
	Forward.	Return.	Forward.	For'w'rd						
100	0	65	0	....	0	....	0	....	0	....
200	65	......	69	....	105	....	116	....	113	....
400	212	291	220	....	291	....	355	....	349	....
600	391	......	360	....	455	....	630	....	600	....
800	529	571	498	....	620	....	918	....	840	....
1,000	663	......	626	....	810	....	1,214	0	1,229	....
1,200	806	853	775	....	1,011	....	......	....	1,539	0
1,400	948	......	918	....	1,234	....	......	....	......	....
1,500	......	......	......	....	......	....	......	10	......	5
1,600	1,090	1,118	1,050	....	1,428	....				....
1,800	1,239	......	1,199	....	1,593	....				....
2,000	1,385	1,385	1,340	0	1,731	0				14
2,500				7	......	6				25
3,000				12	......	11	Failed at a large knot.		Failed simultaneously at two sections.	37
3,500	In this test-piece the central portion was pulled through the knot, so that its tensile strength exceeded the breaking weight.			17	......	18				50
4,000				22	......	24				60
4,500				30	......	32				
5,000				37	......	41				
5,500				42	......	50				
6,000				50	......	56				
6,500				56	......	63				
7,000				61	......	76				
7,500				70	......	82				
8,000				85	......	92				
Total breaking weights in lbs.	8,240				8,100	....	1,830		4,480	
Br'king wgt. in lbs. per sq. in.	11,565				11,095	....	2,485		6,157	
Coefficient of elasticity in lbs.	2,005,050				1,336,300	....	916,640		923,890	
Time of test in minutes.	44				35	....	14		27	

Results of tension tests on specimens cut of an old Douglas Fir stringer, Beam XXII., and of the repeated loading of another specimen cut out of the same Beam. (Fig. 127.)

Loads in lbs.	Readings taken by 1				2		3		4	
	Extensometer.			Rule	Extr.	Rule	Extr.	Rule	Extr.	Rule
	Forward.	Return.	Forward.							
100	0	......	......	....	0	....	0	....	0	....
200	79	141	141	....	117	....	90	....	60	....
400	231	291	292	....	289	....	230	....	190	....
600	389	440	439	....	416	....	376	....	319	....
800	539	580	579	....	518	....	518	....	450	....
1,000	690	730	723	....	635	....	649	....	588	....
1,200	872	872	881	....	765	....	801	....	713	....
1,400			1,030	....	895	....	929	....	847	....
1,600			1,164	....	1,023	....	1,077	....	920	....
1,800			1,340	0	1,169	....	1,205	0	1,096	....
2,000				2	1,304	0		2	1,220	0
2,500				9		8		5		4
3,000				13		13		9		9
3,500				20		18		16		13
4,000				24		23		23		19
4,500				30		29		30		23
5,000				40		36		36		28
5,500				46		42		42		33
6,000				50		49		48		39
6,500				55		56		54		45
7,000				60		62		60		51
7,500				67		71		68		57
8,000				72		83		75		63
8,500				80		86				70
9,000						90				78
9,500						98				
Total breaking weight in lbs.	8,800				10,000		8,520		9,340	
Br'king wgt. in lbs. per sq. in.	12,115				13,954		11,414		13,169	
Co-efficient of elasticity in lbs.	2,139,200				2,199,700		1,969,900		2,190,350	
Time of test in minutes.	17				18		14		14	

Results of tension tests on specimens cut out of Old Spruce stringer, Beam LVII. (Fig. 128.)

Loads in lbs.	Readings taken by									
	1 Extr.	Itale	2 Extr.	Itale	3 Extr.	Itale	4 Extr.	Itale	5 Extr.	Itale
100	0	0	0	0	0	0	0	0	0	0
200	132		130		109		100		75	
400	362		376		286		263		220	
600	614		603		455		431		369	
800	855		843		619		640		525	
1,000	1,121		1,071		834		817		678	
1,200	1,442		1,303		1,017		1,022		829	
1,400		0		0	1,060		1,169		979	
1,500		7		8		9		0	1,124	0
1,600							1,356		1,252	2
1,800								8		8
2,000		19		18	1,239	7		13		13
2,500		32		29		12		20		20
3,000		45		39		20		29		27
3,500		57		50		29		39		32
4,000		69		62		39		38		39
4,500		82		76		49		46		44
5,000		99		89		60				56
5,500				105		71				61
6,000						81				72
6,500						92				80
7,000										88
7,500										96
8,000										
8,500										
Total breaking weight in lbs.	5,500		5,700		6,830 Failed at a small pin-knot.		5,600		7,080		9,000
Breaking weight in lbs. per sq. in.	7,662		7,941		9,361		7,739		10,175		12,626
Co-efficient of elasticity in lbs.	1,032,050		1,202,350		1,025,850		1,063,350		1,577,900		1,903,200
Time of test in minutes.	18		17		16		17		18		16		20

98

Results of tension tests on specimens cut out of Old Spruce stringer, Beam LX. (Fig. 129.)

Loads in lbs.	Readings taken by					
	5 Extr.	Rule	6 Extr.	Rule	8 Extr.	Rule
100	0	..	0	..	0	
200	51	..	127	..	90	
400	191	..	276	..	259	
600	344	..	468	..	445	
800	497	..	652	..	610	
1,000	657	..	870	..	780	
1,100	....	..	960	0		
1,200	811	..	....	..	950	
1,300	....	..	....	..	1,040	0
1,400	967					
1,500	1,040	0	....	5		
1,600	....		....		....	5
1,800	....	5	....	..	....	8
1.900	....	..	....	11		
2,000	....	9	....	..	....	11
2,300	....	..	....	18		
2,400	....	14	....	..	....	17
2,700	....	..	....	25		
2,800	....	20	....	..	....	23
3,100	....	..	....	31		
3,200	....	25	....	..	....	29
3,500	....	..	....	37		
3,600	....	31	....	..	....	35
3,900	....	..	....	45		
4,000	....	35	....	..	....	41
4,300	....	..	....	50		
4,400	....	40	....	..	....	49
4,700	....	..	....	57		
4,800	....	45	....	..	....	54
5,000	....	50	....	61	....	57
5,400	....		....		....	63
5,500	....		....	70		
6,000	....		....	80		
6,500	....		....	88		
Total breaking weight in lbs.	8,100		6,750		5,600	
Breaking weight in lbs. p. sq. in.	11,445		10,206		8,004	
Co-efficient of elasticity in lbs.	1,830,650		1,517,350		1,647,150	
Time of test in minutes.	22		31		22	

99

Results of tension tests on specimens cut out of Old Spruce Beam LXI, and of repeatedly loading another specimen cut out of same Beam. (Fig. 130.) Readings taken by

Loads in lbs.	Extensometer—						Extr.	Rule	Extr.	Rule	Extr.	Rule	Extr.	Rule
	For-ward	Re-turn	For-ward	Re-turn	For-ward	Re-turn	2		1		3		5	
100	0						0		0		0		0	
200	76						91		82		70		79	
400	224	265	255	263	293	276	233		226		198		242	
600	353						289		272		338		413	
800	492	530	526	535	539	548	323		522		475		570	
1,000	631						671		679		613		729	
1,200	774	801	801	807	834	821	819		815		759		881	
1,400	913						964		993		892		1030	
1,500	1051	1051	1074	1074	1085	1055 1095	1100		1173		1029		1198	
1,800							0	0	0	0	0	0	1321	0
2,000							7						1475	6
2,400							11	5	7		4			11
2,500											11			
2,800							17	12	13		15			19
3,000														
3,400							22	18	19		22			25
3,500														
3,800							27	23	25		27			31
4,000														
4,400							32	29	31		33			38
4,800							38	34	37		38			48
5,000														
5,400							44	39	43		44			
5,500														
5,800							50	45	50		52			57
6,000							56	50	57		53			
6,400							61		64					61
6,500							67							
7,000							73							
7,500							80							
8,000														
8,500														
9,000														

Total breaking w't in lbs. 8,980   6,349   6,640   6,300   7,000
Br'k'g w'ht in lbs. p. sq.in. 12,792   9,157   9,724   9,981   9,905
Co-effi'y of elast'y in lbs. 2,866,250   1,999,050   1,551,50   2,070,600   1,826,300

Results of tension-tests on specimens cut out of a 2 in. × 4 in. Red Pine scantling, and also of the repeated loading of another specimen cut out of same scantling. (Fig. 131.)

Readings taken by

Loads in lbs.	Extr. Forward.	Rule	Extr. Forward.	Extr. Return.	Extr. Forward.	Rule Forward.	Extr.	Rule
100	0	....	0	23	0	....	00	....
200	60	....	58	....	55	....	56	....
400	190	....	179	187	173	....	182	....
600	311	....	286	....	279	....	306	....
800	432	....	391	401	396	....	433	....
1,000	553	....	495	....	492	....	559	....
1,200	678	....	600	614	599	....	682	....
1,400	804	....	708	....	712	....	812	....
1,600	929	....	816	837	816	....	942	....
1,800	1053	....	927	....	925	....	1074	....
2,000	1179	....	1035	1045	1039	....	1202	....
2,200	1306	....	1143	....	1142	....	1335	....
2,400	1429	0	1257	1257	1257	0	1461	0
3,000	....	5	....		....	5	....	6
3,500	....	12	....		....	10	....	12
4,000	....	18	....		....	14	....	18
4,500	....	24	....		....	19	....	22
5,000	....	28	....		....	23	....	28
5,500	....	30	....		....	29	....	33
6,000	....	35	....		....	33	....	40
6,500	....	41	....		....	39	....	45
7,000	....	49	....		....	43	....	50
7,500	....	52	....		....	50	....	55
8,000	....	57	....		....	52	....	60
8,500	....	62	....		....	60	....	69
9,000	....				....	62	....	74
9,500	....							
Total brk'g weight in lbs.	9,000		9,280				9,500	
Breaking weight in lbs. per sq. in.	12,689		12,775				14,372	
Co-efficient in elasticity in lbs.	2,279,850		2,554,150				2,247,350	
Time of test in min.	24		20				30	

Results of testing specimens cut out of White Pine Beam, and of repeatedly loading other specimens cut out of same Beam. (Fig. 131A.)

Specimen.

	3							7				8		6					4			5
	Extr.			Extensometer.							Extr.		Extensometer.			time.	Extr.	Extensometer.			Extr.
100	0											0					0	0				0
200	76											86		90			10	80	111			91
300	239		266	278	272	285	296	299				253		220	233	233	17	249	336	336	336	255
400	409	241	266	218								419		305			24	420	544			410
500	579	405										581		630			30	541	752	745	748	572
800	748	569	590	603	600	613	610	629				749		715	564	563	39	739	918			733
1,000	914	732										912		879			48	911	1,136	1,166	1,158	894
1,200	1,082	899										1,076					53	1,104				1,055
1,400	1,209	1,069	1,069	1,083	1,096	1,096	1,109	1,118				1,238		1,045	1,045	1,051	60	1,283				
1,600								1,286								1,223	66		This specimen failed at two section simultaneously.			
1,800																1,396	73					
																81					
																90					

Total breaking load in lbs.	8,260	7,440		9,136	8,470	7,440	6,000	8,696
Break'g load in lbs. per sq. in.	12,252	11,128		12,969	11,561	10,347	9,503	11,981
Co-efficient of elasticity in lbs.	1,835,100	1,799,100		1,729,400	1,654,500	1,614,000	1,123,350	1,741,400

Results of repeatedly loading specimens 2, 8 and 9 cut out of White Pine Beam XLVIII. (Fig. 131A.)

Specimen.

	\multicolumn{4}{c	}{Measurements taken by Extensometer}									
	2				9				8				
100	0	...	...	...	0	...	...	...	0	...	...	...	...	...
200	92	...	...	...	78	...	...	...	92	...	...	...	...	...
300	265	274	...	...	232	231	231	...	234	285	292	324	340	340	340
400	420	...	274	274	392	...	...	256	423	...	...	...	...	...	...
500	583	603	606	593	549	565	564	570	591	613	619	663	648	671	672	677
600	719	...	...	...	705	...	...	...	760	...	...	...	...	...	...
1,000	912	...	...	...	865	...	...	...	929	...	...	...	...	...	...
1,200	1,078	1,078	1,079	1,083	1,027	1,027	1,030	1,031	1,098	1,098	1,102*	1,102	1,150	1,150	1,158	1,158	1,162
1,400	...	...	...	1,110	...	...	...	1,192	...	...	...	...	...	...	...
1,600															
1,800															
Total breaking load in lbs.	6,810				8,316				9,624					
Break'g load in lbs per sq in.	9,321				11,624				14,273					
Co-efficient of elasticity in lbs	1,676,200				1,758,250				1,757,250					

Specimens 2 and 8 failed at two sections simultaneously. Specimen 8, after the reading indicated by a *, was allowed to rest under the minimum load of 400 lbs. for an interval of 2¼ hours. When the loading was resumed the reading was .00324 in.

Results of testing specimens 1 and 2 cut out of Red Pine Beam XXXI, and of repeatedly loading specimens 2 and 3 cut out of same Beam. (Fig. 121 n.)

	1	2	2	1	2				2				3		
100	0	0	0	0	0				0				0		
200	77	99	117	61	92				71						
400	215	293	351	196	290	329	327	340	220	266	266	277	277	280	290
600	319	478	571	313	489				359						
800	504	664	793	445	689	713	518	723	501	541	535	551	547	561	552
1,000	648	854	1,005	571	887				639						
1,200	788	949	1,229	695	1,087	1,087	1,102	1,113	746						
1,400	928	245		828		1,102			937						
1,600	1,067			956											
1,800				1,084					1,086	1,086	1,096	1,096	1,109	1,109	1,117
Total breaking weight in lbs.	8,460	6,928	4,620	7,910	5,392				6,790						
Break'g wgt. in lbs. per sq. in.	11,825	9,379	6,274	10,889	8,090				9,508						
Coefficient of elasticity in lbs.	1,960,500	1,421,900	1,237,500	2,158,800	1,452,200				1,953,100						

SHEARING STRENGTH.

In the experiments, to determine the shearing strength of timbers, considerable difficulty was found in preparing suitable test-pieces which would not at the same time be liable to a large bending action. Blocks were prepared as shown by sketches A, B and C; but unless the sides were sufficiently strongly clamped, as in Fig. A, the specimens almost invariably opened at M, under an effect chiefly due to bending. The clamping, again, introduced a compression, which rendered it impossible to obtain the true shearing stress.

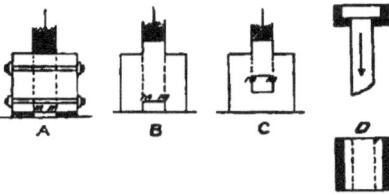

After a number of experiments, more satisfactory and reliable results were obtained by preparing test-pieces as shown by Figs. E and D. The bending action is by no means eliminated, and, generally speaking, it is practically impossible to frame timber joints subjected to a pure shear only. The shearing strengths, which are of importance, are the resistances along planes tangential and radial to the annular rings. An examination of the test-pieces shows that the shears are invariably along these planes.

Thus it will be observed that in the tangential shears, the fibre, both hard and soft, is sheared radially, in the radial shears tangentially, and invariably through the soft fibre.

With test-pieces of the form shown by Fig. D, the shearing strengths along the tangential and radial planes are obtained, while the compound shearing strength, which may be considered as the resultant of the tangential and radial shears, is obtained with the test-pieces of the form shown by Fig. E.

The following Tables give the results of experiments carried out with test-pieces and holders of the form described :—

TABLE OF THE TANGENTIAL, RADIAL AND COMPOUND SHEARING STRENGTHS OF DOUGLAS FIR SPECIMENS CUT OUT OF THE SAME BEAM.

Specimen.	Shearing stress per sq. in. in a direction tangential to the annular rings.	Specimen.	Shearing stress per sq. in. in a direction at right angles to the annular rings.	Specimen.	Compound shears.
No. 1	553	No. 3	560	*No. 13	471
No. 2	568	No. 5	484	*No. 14	536
No. 4	441	No. 7	544	No. 15	629
No. 6	555	No. 8	480	No. 16	657
No. 10	454	No. 9	436		
No. 11	415	No. 12	480		

TABLE OF THE COMPOUND SHEARING STRENGTHS OF DOUGLAS FIR AND RED PINE SPECIMENS.

	Douglas Fir.		Red Pine.	
Specimen.	Shearing strength per square inch.	Specimen.	Shearing strength per square inch.
No. 1	802 lbs.	No. 1	648 lbs.
No. 2	727 "	No. 2	553 "
No. 3	886 "	No. 3	572 "
No. 4	795 "	No. 4	570 "
No. 5	706 "	No. 5	731 "
No. 6	649 "	No. 6	534 "
No. 7	746 "	No. 7	671 "
No. 8		No. 8	698 "
		No. 9	740 "
		No. 10	757 "

Not being altogether satisfied with these results, as the test-pieces did not seem to be of sufficient size to give results which could be considered of standard practical value, new holders, with spherical seats, were designed, and are shown in Fig. F.

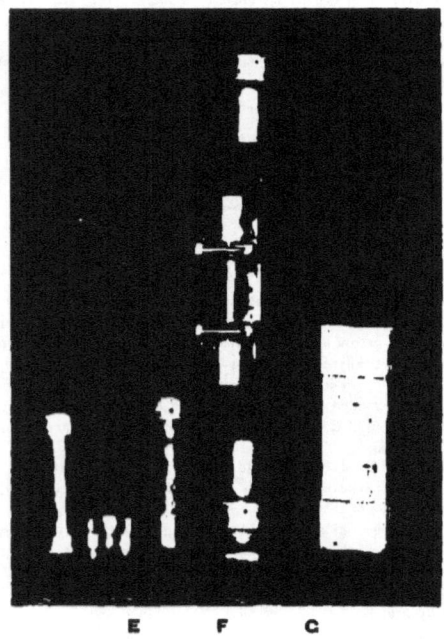

With these holders, tests can now be made upon specimens in which the shearing surface has a width of 8 ins. and a depth limited by the tensile strength of the timber, the maximum shearing area being 96 sq. inches. The web of the specimens is usually about .7 in. in thickness, so that the depth should not exceed .35 t/s, t being the tensile and s the shearing strengths in lbs. per sq. in. The depth of the shoulder forming the bearing for the pressure required to produce the shear is about $\frac{1}{2}$ inch, and is made of only sufficient sectional area to resist failure by compression, as the deeper the shoulder the greater will be the bending action introduced.

From the Tables giving the results of the shearing experiments, the following inferences may be drawn:

  a. The shearing strength of the timbers is much less near the heart than at a distance from the heart.

  b. Generally speaking, the shearing strength increases with the weight per cubic foot.

  c. The shearing strength increases with the density of the annular rings, or rather with the proportion of hard to soft fibre.

  d. A failure sometimes occurs, for which it is difficult to find a complete explanation.

For example, the two specimens from Beam X, and designated in the Table by a *, were precisely similar in dimensions and in weight, and also occupied precisely similar positions relatively to the heart in the stick from which they were cut. One of these specimens failed under a shear of 470.24 lbs. per sq. in., and the other under a shear of 301.84 lbs. per sq. in., so that the shearing strength of the latter was more than 35 per cent. less than that of the stronger specimen. A careful examination of the surfaces of fracture showed no visible difference in the specimens, and the only possible conclusion to be drawn seems to be either that one of the specimens might have been drier than the other, and was therefore deficient in the element of water, or that the shoulders of the weaker specimen, at the end at which the failure occurred, were not cut very parallel with each other, and thus the greater part of the load might have been concentrated on one side.

e. As a result of the experiments, the average shearing strength of Douglas Fir in lbs. per square inch is 411.61, 377.14 or 403.605 according as the plane of shear is tangential, at right angles, or oblique to the annular rings.

In practice, therefore, it will be safe to adopt as the average co-efficients of shearing strength for Douglas Fir. 400 lbs. per sq. inch for shears tangential and oblique to the annular rings, and 375 lbs. per sq. inch for shears at right angles to the annular rings.

Note.—The numbers in brackets at the end of the total shears in the following Table correspond to the numbers in the diagrammatic sections, and indicate the position in the stick from which the specimens are taken. The letter H designates a specimen taken from the heart

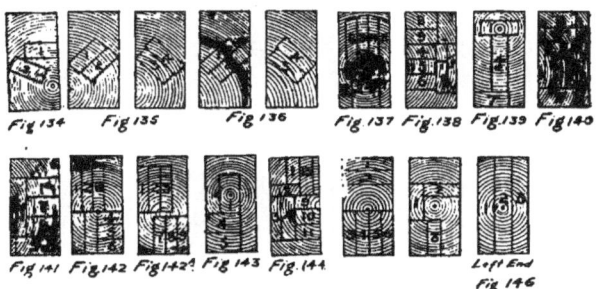

Table of Shearing Strengths in lbs. of specimens cut out of various Beams.

DOUGLAS FIR.

Beam.	Tangential.		Radial.		Oblique.		Av. wght in lbs.
	Total.	Per sq. in.	Total.	Per sq. in.	Total.	Per sq. in.	Per cub. ft.
IX (Fig. 132.)	13,530 (1) 16,610 (1) 16,170 (5) 16,200 (5) 17,210 (1) 16,440 (1)	332·94 404·59 375·47 370·37 412·48 400·09	20,020 (4)	413·40	16,760 (2) 17,120 (2) 14,720 (3) 17,820 (2) 15,820 (2) 17,630 (3) 19,570 (3)	401·22 112·41 393·41 429·05 372·01 360·64 367·99	33·52
	Average 19,380 (2) 15,868 (2) 16,660 (2)	382·65 455·31 477·24 406·14	Average 14,450 (1)	= 413·40 361·23	Average 16,156 (3) 19,430* (1) 12,424* (1) 21,501 (1) 24,880 (3) 23,700 (4)	155·94 = 291·53 170·24 301·84 436·36 311·41 486·29	35·73
X (Fig. 133.)	Average 17,970 (1) 19,760 (1)	= 439·56 433·64 416·51	Average 21,300 (2) 21,300 (2) 16,160 (2) 17,100 (2)	= 361·23 457·30 458·14 317·81 153·79	Average 20,360 (3) 21,500 (1)	133·41 = 398·11 077·67	
XII (Fig. 134.)	Average 16,984 (2) 14,552 (3) 15,330 (4) 15,210 (4) 17,440 (3) 12,940 (4) 12,860 (4) 19,600 (3)	= 425·07 462·15 393·22 414·76 409·97 424·70 443·79 428·80 478·31 432·22	Average 17,686 (1) 16,280 (2) 14,934 (2) 14,920 (1) 15,350 (2) 13,260 (2) 14,610 (2)	= 438·31 464·60 111·04 388·11 355·18 367·07 334·20 350·55	Average	= 137·92	31·57
XIII (Fig. 135.)			Average	= 385·36			31·81

DOUGLAS FIR—*Continued.*

Beam.	Tangential.		Radial.		Oblique.		Av. w'ght in lbs
	Total.	Per sq. in.	Total.	Per sq. in.	Total.	Per sq. in.	Per cub. ft.
XV (Fig. 136.)	19,290 (3)	477·60	15,260 (1)	369·49			36·73
	17,176 (3)	423·00	14,165 (1)	401·50			
	16,170 (4)	420·00	17,914 (2)	431·56			
	16,926 (4)	437·40	16,050 (2)	367·31			
	Average	439·50	Average	397·46			
XVIII (Fig. 137.)	15,272 (14)	446·55			15,495 (7)	359·	
					15,600 (8)	411·9	
					13,120 (9)	447·	
					14,840 (12)	492·5	
					12,595 (13)	402·	
					17,180 (11)	380·	
					12,500 (8)	389·7	
					11,525 (9)	347·2	
					19,120 (10)	392·1	
					Average	400·15	
XIX (Fig. 138.)	Average	446·55	14,430 (4)	375·7	14,470 (5)	393·2	38·4
	16,040 (6)	409·1	14,220 (6)	398·9	20,830 (8)	442·	
	20,390 (7)	422·6	14,590 (7)	411·8	17,200 (9)	371·	
	18,470 (13)	395·7	15,700 (4)	414·6	13,860 (3)	362·7	
	14,650 (13)	340·	15,200 (5)	112·5	15,500 (6)	437·6	
	19,590 (13)	416·5					
	18,865 (7)	410·					
	20,760 (13)	440·8					
	Average	404·90	Average	401·90	Average	401·3	
XX (Fig. 139.)	21,030 (7)	368·3	15,835 (4)	236·7			
	20,635 (7)	413·0	14,270 (1)	252·0			
	21,190 (7)	360·4	17,630 (4)	378·2			
	26,050 (7)	451·1	19,040 (4)	330·6			
	Average	407·0	Average	309·37			
XXI (Fig. 140.)	18,700 (5)	350·	16,840 (2)	291·0	16,050 (1)	282·1	
	17,400 (2)	307·8	14,900 (2)	273·2			
	17,800 (2)	394·	16,560 (3)	307·1			
	Average	350·60	Average	290·43			

OLD DOUGLAS FIR.

Beam.	Tangential.		Radial.		Oblique.		Av. weight in lbs. Per cub. ft.
	Total.	Per sq. in.	Total.	Per sq. in.	Total.	Per sq. in.	
XXII... (Fig. 141)	14,220 (1) 13,370 (5) Average	314· 290· = 302·00	12,175 (7) 14,630 (8) Average	287·0 333·0 = 310·00	17,150 (9) Average	371· = 371·	31·33

RED PINE.

| XXVI... | 20,740 20,850 20,700 18,440 Average | 430·22 (1) 431·67 (1) 366·9 322·3 392·7 | | | 13,020 (H) 16,600 18,680 19,270 Average | 379·59 314·4 347·2 354·2 353·85 | 33·71 |
| From 2 ins. × 4 ins. plank... | | | | | 20,680 (H) 21,900 (H) 18,620 (H) 18,090 (H) Average | 331· 344· 293· 266· = 313·5 | |

WHITE PINE.

| XLVIII... (Fig. 145 and 145A.) | 22,440 (1) 20,665 (2) 16,160 (1) 16,045 (2) Average | 408·89 371·95 430·67 317.96 = 382·37 | 12,120 (7) 11,650 (7) Average | 270·69 275·30 = 272·99 | 14,300 (3) 14,220 (5) 18,505 (6) Average | 364·80 373·89 352·35 = 363·68 | 31·53 |

OLD SPRUCE.

LVII... (Fig. 142A.)	12,100 (6)	386·87	12,975 (3) 11,390 (9) Average	448·96 408·88 = 428·92	8,140 (4) 9,230 (7) 16,075 (2) 13,200 (9) 12,480 (5) Average	403·05 447·85 404·64 457·84 456·59 322·00 415·33	28·37
LX... (Fig. 142.)	16,650 (4) 14,250 (5) 16,400 (4) Average	386·87 302·7 345·4 297·4 = 315·16			17,130 (1) 16,830 (3) Average	292· 283·	
LXI... (Fig. 144.)	13,100 (3) Average =	329·1 329·1	14,800 (12) 14,840 (10) 12,470 (9) Average	460·73 314·6 312· = 362·44	14,000 (12) 12,820 (11) 13,460 (2) Average	287·5 436·78 229·1 404·64 380·17	28·6

N. B.—I wish to express my acknowledgment of the help given to me by Mr. C. B. Smith, Ma.E., in carrying out many of the experiments and in checking the calculations. I have also been ably assisted by Mr. Withycombe, the foreman of the Laboratories, who has devised many mechanical devices which have greatly facilitated the work.

www.ingramcontent.com/pod-product-compliance
Lightning Source LLC
Chambersburg PA
CBHW020143170426
43199CB00010B/859